Salmon Day

The End of the Beginning for Global Business

Douglas Lamont

CAPSTONE

First Published 1997
Capstone Publishing Limited
Oxford Centre for Innovation
Mill Street
Oxford OX2 0JX
United Kingdom

British Library Cataloguing in Publication Data
A CIP catalogue record for this book is available from the British Library

ISBN 1-900961-19-9

Typeset in 10/14 Zapf Calligraphic by
Sparks Computer Solutions, Oxford
http://www.sparks.co.uk
Printed and bound by
T.J. Press Ltd, Padstow, Cornwall

This book is printed on acid-free paper

Dedication

To the family

Those who came before
Those who are with us today
Those who will be with us tomorrow

Some certainties remain from one millennium to the next

Contents

Preface

Finding a title for this book has not been easy. An early possibility, which certainly had the advantage of novelty, was to call it, quite simply, *The End of Global Business*. This is an incendiary title. Might readers not unwittingly think such a title to be a verdict on global firms, which once seemed destined for worldwide dominance, had come to be challenged in ways that those who framed free trade agreements scarcely anticipated? This book, it must be said at once, is not concerned principally with global firms or with their vicissitudes. Rather, it seeks to treat other themes, perhaps no less consequential and certainly no less stressful for global firms. For these subjects, which have to do with the world of NAFTA as it has evolved since 1994, the title of Part 1 seemed appropriate. 'America Leads the Way' from its heartland through its contiguous neighbors to the wider West carries just that element of certainty that seems to reflect sentiment in the US and Canada and uncertainty in Mexico today.

Whether America helps as 'Europe Finds the Way', whether Europe finds the way by itself, or whether Europe and America together create a businessmen's Europe that surges to the wider East – this common pursuit of Euro-realism is a device for understanding emerging trade and investment relationships across the North Atlantic. What makes these elements less certain for global firms is the unfinished state of the European Union.

What will become of the Euro? The Social Chapter? The common defense policy? When and how many central and eastern nation-states will be admitted to the EU? That union of the West, the east–west borderlands and the East with North America is highly uncertain, but critically important to both Europe and America.

The principles of free trade, economic integration and global investments, though frequently challenged in East Asia, had never been wholly abandoned as 'East Asia Discovers the Way'. The economic maps of the region changed in all sorts of dramatic ways in the 1990s, producing a stalled Japan, an apostate China and new paradigms of economic virtue among countries in South East Asia. We live in the shadow of these remarkable success stories, though this is all too rarely acknowledged, especially in the United States. What makes these elements highly uncertain for global firms is the inability to get started on drafting a long-term compact between East Asia and North America.

This book seeks to ask whether the master stories that circulated so widely in the West about American-owned multinational firms replacing nation-states remain valid. Are we not obliged to rethink what these transient successful firms have done to and for the world? Did they not force nation-states to cluster under the umbrella of free trade agreements so that the latter could obtain the benefits of size, scale and scope – all crucial economies in the fight for national wealth, exports, jobs and technology? If the future of global business remains uncertain, it is not because management is underpaid and labor is downsized. There are deeper fissures for global firms in North America, Europe and East Asia – and they are explored in this book – but it would be foolhardy to suggest that knowing them, or even understanding them, provides a protection against surprise. What can be asked for, however, is a greater sympathy for the complexity of societies that are open to and unhappy about the long reach of the global, largely America life style. America's southern neighbors, Europe's eastern partners and Asia's western strangers may, in the twenty-first, be the engines of change, both by the disorders they produce but also by the ways in which their values, however seemingly different, survive and prosper, and not only in their native lands. Today's preoccupation with exclusionary walls around NAFTA and EU may prove to be as short-lived as atomistic nation-states.

This book *Salmon Day: The End of the Beginning for Global Business* continues the inquiry about how global firms prosper, why they fail and how they can avoid failure in the future. *Winning Worldwide: Strategies for Dominating Global Markets* started this line of inquiry with a rich analysis of successful corporate management styles in an emerging world of competing nation-states. *Salmon Day* shows how nation-states can learn a thing or two from global firms and create successful agglomerations of power through free trade agreements. The world, for all its homogeneity, remains a diverse place, whose differences about approaches to free trade, the independence of global firms and commitments to globalization merit close study. Global firms that fail to understand this simple proposition will suffer the fate of Salmon Day – they will work very hard and in the end die.

A note of thanks must go to Richard Burton, a fine editor and publisher, and his partner Mark Allin, for publishing an updated edition of *Winning Worldwide* and this year's most important new book *Salmon Day*.

Douglas Lamont
Chicago, Illinois USA

Introduction

'The velocity, conclusiveness and irreversibility of the shift is unprecedented ... There is no neutrality in the Digital Revolution. You must become a digital revolutionary or risk losing everything.'[1]

John R. Walter, former Chairman and Chief Executive Officer of R. R. Donnelley & Sons, and former President of AT&T Corporation.

'New technology and globalisation are changing the world. But most of the fashionable thinking about the implications of these changes is wrong.'[2]

Pam Woodall, economics editor of *The Economist.*

Win–win–lose world

Let's begin with the true story of R.R. Donnelley & Sons, an American commercial printer that went digital and global under its former chairman and chief executive officer, John R. Walter. Let's remind ourselves that the firm is a product of the highly competitive sink or swim economy of the US, a form of economic combat being pushed by the US government on allies, friends and unsuspecting neutrals alike. Let's conclude with the loaded question: Should Europe, Japan and the countries of South East Asia follow America into this world of cut-throat competition for global market share?

How do you become a digital revolutionary?

Walter lays it out clearly.[3]

- Bet on a radically different future.
- Drive your organization, industry and customers toward a new interdependence.
- Do it fast.
- Toss out traditional financing strategies, the learning curves and the obsolescence rates.
- Make mistakes, learn from them and quickly apply everything you learn.

What is the evidence?

> *'Overnight, the preferred medium for encyclopedias switched from print to CD-ROM. No more do people pay $2,000 (in US dollars) for a printed encyclopedia. Instead, they buy a CD-ROM encyclopedia for $40, or they get it free with the purchase of personal computer ... [This] is a consequence of a theory of the digital revolution known as the cannibal principle. The theorem holds that integrated circuits absorb the functions of once discrete electronic components, incorporate them into a single new chip and give back those functions free.'*[4]

Who wins, who loses?

According to Walter, 'The digital supplier gains a new revenue stream ... Our customers carry less inventory, reduce warehouse expenses and never have a book go out of print ... [But] scores of traditional suppliers lose everything ... R. R. Donnelley & Sons' printing plant in Crawfordsville, Indiana, [with its] world-class press that once ran day and night, producing sets of encyclopedias that would anchor the bookshelves in family rooms throughout North America, ... is quiet most of the time [and never will] run again on a regular basis ... The results: a win–win–lose world.'[5]

What are the results?

The CD-ROM digital revolution is the end of the beginning of a new competitive era for R.R. Donnelley & Sons. 'Walter oversold … growth prospects,' says Kermit Eck, portfolio manager for Cooke & Belier Inc., holder of 2.2 million shares.[6] Even with 'the promise of a high-tech, high-growth, global future … the company [has to manage] its way through a high-cost, mature present … [in which] margins are less than half the 7% they were running in the late "80s."'[7] According to Jonathan P. Ward, who heads the commercial printing unit, R. R. Donnelley & Sons must increase margins rather than revenue growth.[8] With Walter gone, Donnelley's new management is unwinding his digital bets, especially in Poland, China, Latin America and India, with write-offs of over $600 million (in US dollars) and the sale of subsidiaries outright.

How do you become a global loser?

Although Wall Street was not enthusiastic about Walter's appointment as president of AT&T, Walter was labeled by his supporters as a bold leader who had embraced technology and globalization. Hence, he got the better job at AT&T. However, after he left Donnelley, his strategy for the firm showed signs of failure as both revenue growth and margins went down. The company is looking like roadkill on the highway of globalization.

Who referees success and failure?

The financial markets do almost instantaneously as corporate failures affect the stock price and dividend yield of investors. On the other hand, labor markets are slower to react. Wealthy chief executive officers with high pay, outrageous perks, fabulous stock options and, if they fail, golden parachutes go on to another top job before boards of directors are the wiser. On the other hand, employees get laid off. If they get a another job quickly, it is at lower pay without fringe benefits. If they don't, they find America's social welfare net is tattered beyond repair. They too become roadkill in America's quest for higher levels of competitiveness worldwide.

What is the meaning?

America's approach to global success is painfully different to the way executives are paid and redundant workers are taken care of in Europe and East Asia. Today, America's market-rationale approach to investment at home, use of new technology and globalization worldwide is running ahead of attempts by others to compete effectively in global markets. For example, Europe cannot create a European Union-wide corporate tax system, install a functioning EU currency, and bring down industrial labor costs, especially in Germany and northern Europe. Japan cannot break the ties between government and business, give up *keiretsu*, and master the asset deflation created by its Ministry of Finance. Some American economists believe Japan's plan-rationale approach has gone haywire, and the best thing for Japan is to follow the American way of doing business. Others who are social ecologists (the self-described management sage, Peter Drucker) and journalists (for example, Patrick Smith) say Japan is going through one of its periodic reinterpretations of what it means to be Japanese, and how they should live with one another and relate to the outside world.[9]

Is success permanent or temporary?

Is America's number one position in the race towards globalization a permanent feature of economic reform or a cyclical result of current business conditions? Will Japan always be number two, Germany–Europe number three, all of South East Asia number four, and China always knocking on the world's door? Economists are divided in their response. If the answer is yes, then all other countries must follow America's example. Accept the discipline of financial markets. Privatize government services. Let firms sink or swim. And abandon labor, fringe benefits and safety nets. If the answer is no, then all of America's pain is in vain because Europe, Japan and other countries in the world will come back strong to challenge the US.

In summary, R.R. Donnelley & Sons tried to go digital and global, but its senior executives made mistakes in forecasting sales growth and falling margins. America's sink or swim economy worked its will on the firm, and nothing it did overseas could soften the blow. Wall Street punished the firm by pushing down the stock price. New management tried to get out from under globalization without realizing that many roads (including the American approach towards killer competition) lead to success or failure in global markets. In short, globalization is inevitable, but the way individual firms succeed or fail in global markets depends on societal pressures, corporate culture and managerial skill. The trick is getting the right balance. During the 1990s, R.R. Donnelley & Sons spent a lot of time swimming upstream only to get screwed in the end. Let's call this Salmon Day. Although computer jargon, Salmon Day is the appropriate metaphor for the impact of new technology and globalization on markets throughout the world.

Hard lessons

Here's a glimpse of my view about how new technology and globalization are changing business firms. Managers must learn to fire first in several similar global markets all at once. Then aim with some specific market share information. And fire again at crucial segments and niches. Executives must continually answer the key question of their customers: 'What do I get?' For both suppliers and customers today's world is one of short-cycle times, fast paradigm shifts and swift decision-making under uncertainty. It's chaos out there in world markets. Global firms do business in a nonlinear world with no limited number of outcomes, and in which tiny events lead to fundamental changes. Therefore, decision-making under uncertainty calls for managers to employ price-driven cost strategies to gain market share quickly before competitors realize what is going on in terms of price, quality and value added.

Let's look at some of the hard lessons facing digital revolutionaries in a world in which national markets are both local and global all at once, but sometimes in sequence or alternating between the expansion of government on one side and globalization on the other.

- *Lesson no. 1.* Not everyone benefits from the digital revolution. Poorly managed firms encounter lower sales and no profits, and employees suffer no jobs, no income, and no future. Better managed firms mothball redundant factories (such as Crawfordsville), relocate employees and become global digital revolutionaries. The best managed firms learn the trick of getting the right balance between technology and globalization.

- *Lesson no. 2.* Nor does everyone benefit from globalization. According to mainstream economists, global economic integration hurts owners of a country's relatively scarce resources – for example, low-skilled workers in the US.[10] According to pop economists, the loss of work forces government to follow protectionist policies and raise up barriers to international trade among nations, the so-called strategic or new trade theory.

As a consequence, displaced managers, out-of-work employees, disgruntled spouses, and unhappy local retailers clamor for government to do something. 'Protect our jobs,' they say as if they own the jobs for life. For example, the folks in Crawfordsville, Indiana clamored for the state government to save local jobs, give more training assistance to local workers, and protect local suppliers. However, the US federal government only could offer temporary relief for those out of work, search for alternative uses for the almost closed printing plant, and use loans from the US Small Business Administration to keep local retailers afloat.

But the US Congress could keep Chile out of the North American Free Trade Agreement (NAFTA), and it did so during the first Clinton administration. This is a placebo because it does not create jobs in Crawfordsville. For all of these workers it is the beginning of the end of the world as they knew it.

Reimaging

Some refused to learn and sought refuge in an expedient formula called reimaging. Mercantilism is the old-fashioned, eighteenth-century word for government protecting domestic markets from the onslaught of comparative advantage, free trade and competitive enterprise capitalism. In this century, protectionism (or state-sponsored intervention) became an acceptable word among important government economists. For example,

Laura Tyson (then chairman of President Clinton's Council of Economic Advisors) argues in her 1993 book *Who's Bashing Whom*? that 'preventing further erosion of [the] relative economic position' of the U.S. depends on 'activist' government trade policies. Bad policies became good deeds among the polite company of the politically correct in the US.

Lester Thurow (MIT), Robert Reich (the secretary of labor in the first Clinton Administration), James Fallows (then Washington editor of *The Atlantic Monthly* and now managing editor of *U.S. News & World Report*) and Robert Kuttner (coeditor of *The American Prospect*) argue that the new global economy requires a new economic mind set, one that encourages a frontal attack on the economic policies of Japan and Europe. They want the US to become a bit more like Japan, and defend the national economy (and the jobs and wages of American workers) against foreign competition. These pop economists dressed up protectionism, encouraged astute policy engineering by the government, and called it 'strategic trade policy' or the 'new trade theory.'

Notwithstanding the new image Paul Krugman (Stanford University) lambasts the theoretical and empirical errors of pop economists in his 1996 book *Pop Internationalism*. For example:[11]

- The application of the theory of competition among firms to competition among nations is wrong. Firms compete for market share; some gain and some lose; the result is a zero-sum game. When nations use comparative advantage to carry out international trade and investment both exporting and importing nations gain, although one may gain more than the other (or as economists say, an asymmetrical gain). The result is a positive-sum gain.
- The assumption that Third World economic growth harms the economies of the US, Europe and Japan is wrong. The 'tigers' and the 'tiger cubs' of South East Asia, Latin America and eastern Europe are all recent examples of how an expanded role for the Third World has helped the economies of the developed world.
- The use of preferential government policies to protect (stimulate or grow) high-tech industries is wrong. Their impact on the growth of the gross domestic product and increases in real wages is negligible and may in fact inhibit growth altogether.

These three popular misconceptions send the wrong messages out to the world. Moreover, the 'facts' upon which they are based are wrong, too. The US is still not so dependent on the global economy as the new trade theorists suggest. The statement that the extent of 'globalization' is historically unprecedented is also wrong.[12] 'On some measures, economies at the turn of the century were every bit as open and integrated as they are today. Most industrial countries' trade as a share of GDP is not much larger now than it was a century ago.'[13] US dependence on international trade has not changed much since the 1970s where it was and still remains today about 10–11% of gross domestic product.[14]

Turn the facts around. American firms sell about 90 percent of their output to their own workers in the US. 'Americans do not have to do anything or change anything to be part of the global marketplace,' says Murray Weidenbaum, the Mallinckrodt Distinguished University Professor and chairman of the Center for the Study of American Business at Washington University in St Louis. 'Even if a business does not export a thing and has no overseas locations, owners, managers, and employees are still part of the world economy ... The issue has been decided by technology. The combination of fax machines, universal telephone service (including cellular), low-cost, high-speed copiers and computers, and speedy jet airline service enables money, goods, services, and people to cross most borders rapidly and often instantly. And that goes especially for what is the most strategic resource – information.'[15]

Intellectual revolution

Management sage, Peter Drucker, says 'The 20th century has made ideas mobile'.[16] Intellectual information is spreading so fast – via the personal computer, the Internet and the digital revolution – that it is upsetting lives of bureaucrats, business executives and people. Information demystifies governments, businesses and chief executive officers, exacerbating the loss of confidence in them. Even 'the bazaar, teeming with sleek young things from [London, Frankfurt, Tokyo and] Wall Street, has lost its mystery'.[17] For example, Drucker forecasts an end to industrial organizations (such as commercial banking) and the rise of alliances in which individuals with crucial information act as design consultants to those without good intellectual ideas.[18] The former cluster around loose informal organizations, and create new businesses, some far removed from customers. Hidden in these changes are Drucker's crucial questions: 'What needs to be done? What do I get? What can I do? Tell me what you are going to do on Monday that's different.'[19]

John S. Reed, the chairman of Citicorp, counters, and suggests wisdom and understanding come from thinking about the interactions and the tensions among four markets:[20]

- The *information* market, which seems almost perfectly free.
- The markets for *goods* and *tradable services* – significantly liberalized but still characterized by a fair degree of protectionism.
- The market for *labor*, still comparmentalized by nation-state.
- The *capital* markets that are quite open but still function with some stickiness and discontinuities.

Comparative advantage

Notwithstanding these visible tensions, the US does gain from international competition by specializing according to its comparative advantage. Today, in the last few years of the twentieth century, the US has a comparative advantage in the digital revolution *vis à vis* the nation-states of the European Union and Japan. Although the US makes CD-ROMs more efficiently than Europe or Japan, both might make CD-ROMs whenever their domestic wages are cheaper than in the US or when the foreign exchange value of their output (denoted in Deutschmarks or yen) is cheaper than American output expressed in US dollars. So international trade raises incomes in both Europe and Japan, and in the US. Foreign trade is not a zero sum game for nation-states.

However, income does not rise as rapidly as many economists think. There is a lot of intra-industry trade. Sometimes foreign suppliers encounter lower wages in importing countries. More often complete specialization does not normally occur, notwithstanding what economists believe about comparative advantage. All we can say is that international trade raises incomes in the aggregate, which makes it possible for losers to be compensated with the net gains of others.

John Walter as the new president of AT&T Corporation, R. R. Donnelley & Sons as the new manufacturer of CD-ROMs, and the US as the new source of the digital revolution are all trying to benefit from new technology and economic specialization. These are the asymmetrical benefits of wage freedom, enterprise capitalism and comparative advantage. All can be enhanced by going international and doing global business, or what we now call 'globalization.'

No place to hide

According to Raymond Vernon, Emeritus Professor of International Affairs at Harvard's Kennedy School, 'People don't invest overseas because they're happy about it. They do it because they're scared. It isn't as if they've got a hell of a lot of choice. There's no place to hide today for many, many industries. But the main message is to prepare yourself for a headache.'[21]

Karl Sauvant, research director of a division of the UN Council on Trade and Development in Geneva, Switzerland, points out that many multinational firms want to 'globalize production' to become more competitive.[22] For example, the UN reports in 1995 multinational firms invested a record $315 billion (in US dollars) to acquire competitors and make 'greenfield' investments. 'Cross-border acquisitions in the pharmaceutical, telecommunications, financial-services and entertainment industries spurred the big jump in foreign investments.'[23] The US received $74.7 billion in foreign direct investments and made $49.4 billion in 1994 to other countries. In 1996, these include investments by all industrialized nations in the following countries: China received $52 billion, Mexico $28.1 billion, Indonesia $17.9 billion, Malaysia $16 billion, Brazil $14.7 billion, Thailand $13.3 billion, Argentina $11.3 billion, India $8.0 billion, Turkey $4.7 billion, Chile $4.6 billion, Russia $3.6 billion, and Hungary $2.5 billion in foreign direct investments.[24]

'The trend among companies is to get closer to their customers and become a major presence in the markets where they sell,' says Robert

Hormats, a vice chairman of Goldman Sachs (International). 'Countries all want foreign investment. The old porcupinish days of nationalism haven't gone away but countries increasingly realize they need to have an appealing [foreign direct investment] environment to attract jobs.'[25]

'What's emerging is truly a global economy,' says David Mulford, chairman of CSS First Boston Ltd. 'It used to be that a large U.S. company had investments all over the world. Now if a company like that buys a European company of the same size it becomes a very different situation' that gives the parent company more of a global foothold, he adds.[26] The growth in foreign direct investments stems from competitive pressures, new technologies, privatizations of country-owned enterprises and more of an open-door policy toward investments by many countries. The combination of factors means many companies 'no longer can produce goods at home and move them out overseas; they have to be producing and making the key strategic decisions [overseas] as well,' says Sauvant.[27]

Winning worldwide

In 1991 I wrote the following about some of the futures that have already happened in the continuing battle for international competitiveness:[28] 'A global international management consciousness exists among rival firms irrespective of their country of origin. This spills over into standardized product design concepts that are erasing local and regional identities as consumer preferences converge worldwide.[29] However, varieties in national and corporate culture require firms to exhibit flexibility in pursuing customers while at the same time pushing for global integration. The dynamics force firms away from the commodity nexus of export-import trade and the power nexus of direct investment and ownership by multinationals toward partnerships, joint ventures, research and marketing consortia, and so on.[30] This radical restructuring is required of firms that wish to sustain their international competitiveness in the face of changes in the political economy of the world.'

Foreign direct investment, especially in countries in which local competitors are tough and resourceful, is not a no-brainer. The American firm, Whirlpool, learned this lesson the hard way. In 1989, 'Whirlpool paid $960 million (in US dollars) for the appliance unit of Dutch electronics giant Philips and spent another $500 million to retool its plants ... [But] Whrilpool's European revenues remain flat ... since 1990.'[31] Here are the reasons for Whirlpool's failure in Europe:[32]

- Americans buy on price and throw away old appliances. Europeans think of appliances as long-term investments, and the latter expect more in finish, durability and appearance.
- Americans put their washers/dryers in garages or basements where noise is not noticed. Europeans put their appliances in kitchens where noise does matter.
- Whirlpool reorganized its sales force, which turned off European customers.
- Whirlpool sold cheaper models and made less per unit than its competitors.

Of course, strong local competitors, such as Bosch–Siemens and Electrolux, took advantage of Whirlpool's mistakes. See Chapter 10 of my book *Winning Worldwide* for how Electrolux beat back the European challenge of two American appliance manufacturers, Whirlpool and Maytag.

In 1994–5, Maytag's new chief executive officer, Leonard Hadley, bit the bullet, and sold off the firm's Australian and European business. Maytag dropped its sales overseas from 21% to 6%, cut its long-term debt by 38%, and pushed its returns on equity and capital – 18% and 12%, respectively – to twice that of Whirlpool. Instead, Maytag decided to focus in overseas deals on a few carefully picked alliances – for example, a joint venture with Hefei Rongshida, a Chinese appliance maker that produces 10% of all washers in China.[33] This new strategy about global markets seems to be working.

In 1996 I added the following comment to the second edition of the same book:[34] 'Because global firms are the dominant economic institution in the international economy of the 1990s their future has several important cautions for the dominant political institution in the world, the nation-state. These are the warning notices.

'Executives are better able today than anytime in the past to manage country risk. They avoid governments, peoples, and nations where tradedefying practices make up the preponderance of government policy toward business firms. Japan, the United States, and Europe are exceptions because of their large internal markets. Canada and other small, developed and less-developed countries must conform to what executives expect from governments or face the loss of revenues from future foreign direct investments. Because executives live within an international management culture that transcends national political differences and one that looks for new ways to link up with others across industries, nations, and continents, global firms will continue to be more powerful than most nation-states irrespective of resources, macro economic policies or Nobel peace prize winners.

'Also executives are more competent today than anytime in the past managing product sourcing, customer selection, and value creation. They have internalized within their firms the export platforms created by countries, the high quality consumers trained by educational ministries, and the proprietary information first created by now privatized (but formerly government-owned) laboratories and enterprises. With the end of state socialism in central and eastern Europe, Russia, and the former Soviet Union, these three processes of internalization will quicken until only the most backward regions of the world are left outside the world's market economy.

'Finally, executives seem to be better able than most government officials in planning for the rapid rate of changes in the world economy. Although both national governments and big global firms share the problem of entrenched bureaucracies, the latter seem to understand better the need for organizational learning, if only because financial markets demand increases in sales, profits and shareholder value. Very few national governments can match the success of a Unilever or an Electrolux in globalizing the firm, its products, and markets so that consumers worldwide benefit from the material success of these global firms.

'Although nation-states will be the dominant political institution well into the next century, their role in the world economy and their relationship to transnational firms is evolving from what it used to be to what it might become. The willingness of governments to form com-

mon markets, customs unions, and free trade agreements is a hopeful sign that even nation-states see the benefits of freely competitive, open markets. If executives of global firms argue for free markets first regionally and then worldwide, they will become as active a set of players in world politics and as they are in the world economy. This is the most important task facing international executives as they plan for the future of their firms in the world economy.'

Coke versus Pepsi

Let us see how the principles of *Winning Worldwide* are applied by two rivals in their quest for dominating global markets. Coke is kicking Pepsi's can worldwide. According to Caroline Levy, who follows the soft drink industry for the investment firm Schroder Wertheim, 'Pepsi is losing customers to Coke in every major foreign territory. [Pepsi] has always struggled overseas, but in the past few months it has lost key strongholds in Russia and Venezuela to Coke ... By acquiring 50% of a $400-million-a-year bottling company, Coke put Pepsi out of business there overnight. The coup in Caracas is particularly painful for Pepsi, since the bottler was one of Pepsi's most respected. because Pepsi outsold Coke 4 to 1 before Coke's acquisition of the Venezuelan bottling company ... Also Pepsi's largest bottler, Balsa of Buenos Aires, is essentially bankrupt. Its new plants in Brazil are running at a third of capacity. PepsiCo owns 24% of the imploding operation, which will obliterate Pepsi's international drink profits this year.'[35]

Why is Coke more successful than Pepsi? 'Coke opts for the long run ... with its anchor bottler concept. Coke has eight large, solidly capitalized bottlers around the world; the company owns equity stakes in each and allows them to expand gradually ... Pepsi tries to replicate in three years what Coke has built over 50 years ... [Yet] Coke generates 71% of its revenues in international markets and derives more than 80% of its income abroad; on the other hand, Pepsi generates only 30% of its revenues and 20% of its income from overseas markets.'[36]

Coke is truly global and winning worldwide. Pepsi is not global and losing worldwide.

Globalization

Here's a glimpse of my view about the future of globalization. The US, the EU and Japan are at a crossroads. They and their partners in the Organization for Economic Cooperation and Development (OECD) narrowed the gap in living standards between the richer (let's call them the *haves*) and rapidly developing countries (let's call these the *have-lates*) of East Asia, Latin America and eastern Europe. In fact, some of the *have-lates* have caught up with the *haves*, and the former credit the distributional effects of globalization for their economic success.

However, large parts of the world (let's call them the *future-haves*) were left out of the process, falling behind in terms of living standards, incomes, and wages. Here the distributional effects of globalization are blamed for their lack of economic success. Many countries in South and Central Asia, sub-Saharan Africa, the Andean regions of South America and elsewhere worked hard to change government policies and to offer low-wage incentives to foreign investors.

Most multinational companies prefer to do cross-border deals within other OECD countries rather than risking their foreign direct investments in places with low wage costs, poor manufacturing productivity and inefficient government. According to a study by the Deloitte & Touche Consulting Group, 'access to cheap labour is a rapidly decreasing priority for US manufacturing companies investing abroad, which are chiefly interested in establishing positions in large and prosperous markets with worldclass production standards.'[37] The *future-haves* ran the race of economic development, first with state-sponsored intervention and then through private marketplace initiatives, and came in last behind the *have-lates* and the *haves*; because, the poor countries only have cheap labor to sell and most multinational firms do not want to purchase the output from very low wage areas of the world.

Global business strategy

Let's tie all this together. Here's how Kenneth S. Courtis, chief strategist and economist with Deutsche Bank Capital Markets (Asia), the investment arm of Germany's giant Deutsche Bank), frames his thoughtful questions about global business strategy:

- Does history explain trade and investment among nations?
- Do current events portray key international economic trends?
- Is there a realistic set of futures for global business?

'Ken thinks in three dimensions: historical, current and future. He is a genius in capturing the essence of a trend or a set of forces with language that resonates with both experts and laymen. He has the keenest sense of the interconnection between economic and political forces of anyone I've seen. He's totally unique as a global strategist,'[38] says Jeffrey E. Garten, former Undersecretary of Commerce for International Trade in the first Clinton Administration and now Dean of the School of Management at Yale University.

Courtis's forecast is as follows: 'If global companies a decade from now are not generating one-third of their revenues in Asia, they will no longer be global. They will be large organizations but ones increasingly regional and on the defensive.'[39]

Coke in Asia

Coke is pouring into Asia, especially into 'China, India, and Indonesia, which together have 2.4 billion people, almost half the world's total ... Coke predicts that each of the Big Three Asian markets should double sales roughly every three years for the indefinite future ... The populations of these countries are young, incomes are exploding, and the drink market is undeveloped ... What Coke is really running up against is the need to change the drinking habits and lifestyles of millions of new customers ... Coke has 23% of the soft-drink volume in China and figures on eventually topping 40%.'[40]

Coke's problems in Asia are as follows:

- Coke does not have the control it needs to make merchants follow the rules.

- Infrastructure for the transport and distribution of Coke is primitive.
- Coke has difficulty finding the right local partners with good government contacts.

One problem Coke does not have in Asia is strong competition from Pepsi. In the words of Ken Courtis, Pepsi will be a large organization but one increasingly regional and on the defensive while Coke is already a global company generating increasing revenues in Asia.

According to Morgan Stanley, an American investment bank, Coke has significant competitive advantages in the soft drink industry. These include currency values, relative unit labor costs and technological prowess. Also Coke can sustain these advantages for several years while Pepsi flounders helplessly around the world.[41]

Ford worldwide

In 1994 'Ford Motor Company decided to tear up its management structure, abolish its regional companies and turn itself into a global company that made global cars. Ford 2000 pushes together the company's European business (sales: $23 billion) with its North American one ($105 billion) ... This amounts to a merger of two companies that makes most real takeovers seem puny ... Full-scale globalization remains a difficult trick to pull off ...

'In both America and Europe Ford is facing stagnant markets and resurgent competition form its old rival, GM, as well as from Japanese car transplants. It will take years for Ford's strategy to bear fruit in Asia, particularly as rivals such as GM, Peugeot, Citroën and Volkswagen have established formidable leads in both China and India. Most worrying of all, the reorganisation has neither boosted the quality (and stylishness) of Ford's cars nor – yet – slashed the company's development costs.'[42]

Also Ford's traumas in South America follow the dissolution of Auto Latina, a joint venture launched a decade ago with Volkswagen. This left Ford with a severely weakened presence particularly in Brazil, the largest new car market in South America, and kept this Ford subsidiary outside the Ford 2000 program.

In Japan the task is to include Mazda (Ford owns 33.4%) in the Ford 2000 program. The Hofu plant is Mazda's most modern and efficient plant, but it is also the least flexible.[43] The warning signs are that Mazda gives Ford redundant small-car development operations in Japan and forestalls the integration of Japan into the Ford 2000 program. Ford may have taken on more than it can do well.

Government policy and economic growth

According to Professors Deepak Lai of the University of California at Los Angeles and H. Myint, formerly of the London School of Economics, 'the proximate causes of growth are the rate of investment and its efficiency ... But these in turn are determined by the actions of governments, which can underpin – or demolish – investment opportunities. [Here] are the three things they must do: (1) provide the basic public goods, above all security of life and property, at tolerable cost in terms of taxation; (2) ensure fiscal and monetary stability; and (3) allow citizens to exploit the global economy ... Will governments do better in the future? Possibly, [because most] have a wider appreciation of the role of the market,'[44] and its demand for self-discipline and reform.

On the other side of the argument are Pierre-Richard Agenor, and an economist at the International Monetary Fund, and Peter Montiel, a professor at Oberlin College in Ohio. They belong to a school of economists called development economics, a branch of economic theory that grew up in the 1950s and 1960s 'around the question of how to promote economic development in poor countries ... Poor countries are structurally different from rich ones, and so needed their own set of economic models.

'Developing countries ... tend to be more open than richer ones (that is to say, trade contributes a bigger fraction of national income), and to depend more on foreign capital. They tend to have fixed exchange rates and, often exchange controls. Their financial markets are rudimentary and often distorted by heavy government regulation. The public sector plays a bigger role than in rich countries particularly in directing the pattern of investment.'[45] All that can be said is the poor envy the rich and this gives business executives the opportunity to prosper in global markets.

Envy leads us to the conclusion that the *future-haves* must free their markets and curb the role of the state, or they will not join the ranks of the *have-lates*. The poorer countries do not need their own branch of economics, but just better macroeconomic policies. Here is a short list:

- Dismantle trade barriers.
- Tighten fiscal policy.
- Privatize state-owned firms.
- Attack inflation.
- Create incentives for profits.

Do nation-states follow the American market-rationale approach, the Japanese plan-rationale approach, or some hybrid found in the countries of South East Asia? Whither India and South Asia, Turkey and the Middle East, and South Africa and sub-Saharan Africa?

Let us jettison the talk about the irresistible forces of globalization. According to the *World Investment Report 1996* from the United Nations Conference on Trade and Development, Asia and America is where the smart money is going in the mid-1990s. 'The supposed infusion of rich-country investment in the world's emerging economies ... is happening, but selectively' in the auto assembly plants in Brazil, in natural resources in Chile and relatively little in Africa.[46] 'Sustained economic success depends on establishment of a *competitive* market economy. But this is something that only social *co-operation* can achieve. The wealth and poverty of nations is determined not by impersonal forces, but by political failures ... This, in a nutshell, is the contrast between the [rich] United States and the [poor] Mexico.'[47]

Therefore discontent turned into sales, profits, and business opportunities, competitive advantage, and sustainable development depends on the quality of government. It is a desire to participate equally in the major markets of the world. Global envy is to covet the infra structural resources and the managerial capability that other nations have. It also means joining the struggle to create a worldwide global market economy with the aid of the digital revolution.

Increasing market share

Gaining a larger market share requires decision-making under uncertainty. It involves an entire team of company executives who are determined to develop product line extensions and making sure that products are sourced properly.

Today's global business firm is nonlinear. It has no limited number of outcomes. It does not have a strategic intent. It doesn't do long-term planning. It fires first. Then aims. And fires again. Therefore, tiny events lead to fundamental changes.

Teams of marketing executives must spot emerging competitive threats, create disruptive technologies to counter these threats, apply them rapidly to emerging markets, or mobilize to attack them before they become serious problems. This is 'bounded instability,' and it leads to creativity and changeability. Here's a seven-step outline of what successful marketers do in export markets:

1. Produce chaos.
2. Work with short-cycle times.
3. Carry out a price-driven cost strategy.
4. Add value through customer-based innovations.
5. Mobilize customers to create their own value added.
6. Force their industry into paradigm shifts.
7. Develop customer partnerships in each market segment.

This is our to do list. In short, successful international marketers focus on what they bring (in added value) to other players, and how they can lower the value added of competitors. Successful global executives shape the game by changing the rules. Successful exporters delineate a mixture of co-operation and competition among players. For them global marketing strategy is the stretch between ambition and resources.

So what is different today for exporters, international marketers and global business executives? They too must follow the seven-step outline listed above. They too must search for information about business opportunities abroad. They too face frequent turnovers in the stock of knowledge and they too must act before the acquired information becomes obsolete. That is why the digital revolution is so important. The Internet helps us deal with the *complexity* of modern global business.

So what is different today for exporters, international marketers and global business executives? They too must confront strong vertical relationships among suppliers, distributors and end-users. They too must deal with the expense of switching costs and the refusal of potential buyers to accept new market information (on products, prices and after-market service). They too must deal with the reality that free trade agreements and customs unions reduce opportunities for outsiders to break in and sell new products. These FTAs deal with the *paradox* of modern global business. Today, it is both global and local.

Think global, act local

What is the future of global business? Business executives must pay attention to following uncertainties:

- *Social dynamics.* What will be the demographic makeup of markets in the countries of the *haves, haves-late* and *future-haves*? Will the youth of the first world find more opportunity in less-developed countries? How influential will youth be in 10 years? As adults will they still drive the technological revolution? Will people get bored with the Internet, online chatting and nonpersonal relationships?
- *Economic issues.* How will international trade flow and exchange rates affect the price of computer-based telecommunications, energy, food and manufactured goods? How will governments govern the market? Will competitors form alliances with governments? How might the structure of the industry change as new markets and new sources of supply enter the game of global business?
- *Political issues.* Will governments permit free trade agreements and customs union to continue to undermine national sovereignty? How many new mini-states will the world accept as legitimate aspirations of ethnic, religious and local tribalism? Will the UN gain the ability to tax the 'Fortune 500' firms so that the world's peace-keeping operations can be fully funded before they explode into wars that impede global business?

- *Technological issues.* Will the *haves* and *have-lates* be willing to give up dirty technologies before the environmental impact is irreversible? Can the *future-haves* leapfrog to clean and light technologies? Will this shift widen the gap between rich and poor?

Cultural change

Bernard Lewis opens his history of the Middle East with an Arab Muslim consuming 'Western' goods while saying that 'Western culture' was alien to him and his cultural group. 'In modern times, the dominating factor in the consciousness of most Middle Easterners has been the impact of Europe [first Spain and Venice, then England and France], later of the West [especially America for the last 50 years] more generally, and the transformation – some would say dislocation – which it brought.'[48]

Today, the dominant consumer culture worldwide is the global, largely American technologically based culture. It drives the routine social patterns (RSP) of global consumption, segmentation across similar markets worldwide and targeting products based on market share analysis in key nation-states. Call it the digital revolution, the Internet culture, IBM culture, the Compaq culture. Also call it the NAFTA or European Union free trade culture. Moreover, call it the McDonald's culture, the Levi's culture, the Coca-Cola culture. There is a national culture and then there is the international culture which is mostly American in emphasis, or Chinese with family connections in practice. Most global players hold dual citizenship – passports from nation-states, and network commitments to family and friends, and bureaucrats and other business leaders.

Here's a glimpse of my view about the future of global culture change. First, cultural identity changes over time. Second, the digital revolution is speeding up the process of cultural change, but the impact is uneven across the world, among political, economic and social institutions, and even within free trade areas. For example, the European Commission conducted a poll and 'found that half of the people of Britain, Portugal and Greece thought of themselves either in purely in national terms; so did a third of the Germans, Spaniards and Dutch. And this was in a part of the world where there is an institution – the EU itself – explicitly devoted to the encouragement of "Europeanness".'[49] Third, cultural identity is a bundle of characteristics – that is, a mix of traditional values, and modern rational and scientific thought; of religion, information and scientific knowledge; and a combination of the past facts, present realities and future possibilities.

But I am reminded of what the senior minister, former prime minister of Singapore Lee Kuan Yew said, 'Culture gives coherence to society, the national culture, the regional culture, the local culture is very important.' And this is crucial to doing a good marketing job – tailor products to fit national markets. There are some things which are truly international. There are other things which cross only a few markets and some things which sell all over the world.

In the final analysis, the world market is not a cultural concept.

Salmon Day

Salmon Day is the experience of spending an entire day swimming upstream only to get screwed in the end. R.R. Donnelley, Whirlpool, Maytag and Pepsi fought the global fight, and lost market share to their competitors. Only Maytag has turned itself around to the satisfaction of Wall Street. Although Salmon Day is computer jargon, this is the appropriate metaphor for today's global business.

Part 1 explains America's new political economy, nationalism and ideology. The US is perhaps the only country that can be both a political nationalizer and economic globalizer; because it does not have to use nationalism to manage the societal and economic tensions created by technology, the intellectual revolution and globalization. What will happen to the US (and Canada) as it becomes less European, and more Latin America and East Asian in cultural attitudes and approaches to work? America is going through a transition from what Americans know today to what Americans might become tomorrow. Will America face Salmon Day? Probably not.

Part 2 explores Europe's emerging free trade commitments to eastern Europe and Russia, and to the countries on the other side of the North Atlantic, the NAFTA countries. The European Union can act as a political nationalizer and economic globalizer whenever its 15 member-states permit it to do so. Usually, the EU is dependent on the nationalisms of its nation-states to manage budget deficits, high levels of unemployment and the creation of a common currency, the Euro. All these tensions stem from America's commitment to introducing high-technology, accepting the mobility of ideas and pursuing global market share. What will happen to the EU as it becomes less western European, and more central to creating a common economic entity from California in the west to Poland in the east? Europeans are going through a transition in which the unknowns far outweigh the knowns. Will Europe face Salmon Day? Probably yes.

Part 3 discusses East Asia's commanding presence in the Asia–Pacific region, and whether Japan, China or the countries of South East Asia will take on a leadership role individually or collectively to cement America's ties to East Asia. At least within the region, Japan today and China tomorrow seek to be political nationalizers and economic globalizers. Both externalize their societal and economic tensions through nationalism – in large part against the US. What does East Asia offer the US for it to shift its future from the North Atlantic to the Pacific? Japan, China and South East Asia suggest new paradigms for global business. Will East Asia face Salmon Day? Probably no for Japan and the South East Asian countries. Probably yes for China.

Other casualties of Salmon Day might include Mexico and Central America; the Andean countries of South America; Europe east of Poland, the Czech Republic and Hungary; Russia and the CIS countries of central Asia; South Asia and sub-Saharan Africa. Their governments must respond to unknown probabilities of extremely unpleasant outcomes – that is, do nothing or intervene actively, and still get screwed in the end. Their governments must learn how to deal with others who know more than the former – that is, if nothing is done or if everything is done, others have asymmetric (or private) information and can distort possible good works form developing countries that want to follow the path of the economic tigers of East Asia.

Decision-making under uncertainty has some ideas about how governments might enlist business firms to avoid Salmon Day. More on these in the following pages.

Part 1

America Leads the Way

How to Reimage America's Place in the Global Economy

American Heartland
Nafta and Mercosur
The West

Executive decisions

Remember America has choices. In Chapters 1 through 3, the following choices are discussed. They are divided into two categories. The first set has a high probability of success in avoiding Salmon Day for America, Europe and the West. The second set has a low probability of success in dodging Salmon Day for the America, Europe and the West. You decide what to do.

Decisions leading to success

- The West creates a North Atlantic Free Trade Area by combining NAFTA and the EU.
- The West and the nation-states of East Asia create an Asian–Pacific Free Trade Area.
- The WTO promotes market economies within the Orthodox and Moslem communities.

Decisions leading to failure

- America subsidizes Mexico's efforts to succeed within NAFTA.
- America and Mercosur support the accession of Chile to NAFTA.
- America, Mercosur and the Andean Pact create a Free Trade Area for the Americas.
- Europe extends the EU to eastern Europe, Russia and the former Soviet Union.
- America extends NAFTA-type free trade to the Middle East.

The work of Salmon Day is about making decisions under uncertainty.

Chapter 1

American Heartland

'Culture is not what was done but what is passed on by a complicated cast of "transmitters" – from politically correct pop economists to the media, professors, business executives and international civil servants.'[1]

Gary Taylor, a Shakespeare scholar at the University of Alabama and the author of *Cultural Selection.*

'America's urge to bigness is likely to endure because, once entrenched, it tends to spread and magnify.'[2]

The Economist.

Cultural memory

Memory mothers internationalism, babies globalization, and nurses the interaction of governments, business firms and citizens with free enterprise markets. Also memory coddles the global, largely American life style, and humors its impact on the slow disappearance of cultural diversity around the world.[3] Moreover, memory requires global executives to use three English words, World Wide Web, for their Internet language of communication with far-flung subsidiaries, suppliers, customers and tax authorities.

The Internet only works for you if you have a solid command of the English language. Because whether you are French state enterprise managers pursuing the decline of Colbertism, Japanese salarymen curious about new selling techniques, or Mexican consultants concerned about the availability of venture capital, the World Wide Web really only works as a great unifier if you speak English.

Learning English is 'fast becoming America's greatest and most effective export (carrying with it immense cultural power) ... More people ... speak English as a foreign language than speak it as their first language. English is already the language of diplomacy, scientific discourse, air traffic control.'[4]

What do we remember?

- America's military dominance of the Western world.
- Huge investments by American business firms.
- Management theory by Drucker, Porter, Kotler: all Americans.
- The big assignment on free trade with Mexico, Europe, Asia.
- Three English words: World Wide Web.

And, of course, contemporary American culture: the way we celebrate Christmas, eat at McDonald's, drink Coke, play at Disneyworld, download information over the Internet.

What are we reinterpreting? How America became the first country to make ideas mobile? Where is the intellectual revolution going in America's sink or swim economy? Why are America's cultural memories changing and how will these trends affect others in the world? What it means to Europe and East Asia for America to be both a political nationalizer and an economic globalizer?

Language memory

In 1995, 'at the biennial meeting of *La Francophonie*, in Cotonou, Benin, France's new president, Jacques Chirac, issued a call for French-speakers to stake a claim on the Internet (the *inforoute*, he actually called it), a communications medium that otherwise seems destined to become an English-language monopoly. In making the case for linguistic pluralism and cultural diversity on the *inforoute*, Chirac observed that if even French, which is spoken by some 150 million people – making it number nine in the world, after Chinese, Hindustani, Russian, English, Spanish, Arabic, Bengali, and Portuguese – is faced with uncertain prospects, the situation must be much worse for smaller languages [Navajo, Topoiyo, Inuit, Irish, Flemish], which in Chirac's view risk complete eradication.'[5]

Language harbors valuable conceptual resources. American English captures the diverse resources of political empire, corporate power, management practice, marketing techniques, Hollywood and the Internet across all continents in the world. Chinese secures East Asia for wealthy overseas and mainland Chinese families and plants its flag in North America, too. Hindustani, Russian, Arabic and Bengali are tied to ancient homelands and their 'near abroad' conquests. Spanish, Portuguese and French make the hard journey to new lands and prosper creating new native speakers. However, none of the ten most used languages in the world compare to English's world wide acceptability.

Only Spanish tries to run the race with English in North America, and it too fails as the American-born Mexican middle class accepts the cultural memory of its new homeland. Within two or three generations, the cultural memory of old Mexico recedes into a distant past as the children speak English first and Spanish second or not at all. America did this to Germans, Russians, Italians, Poles, Czechs, Greeks, Japanese, Chinese, Thais and many other people of the old world. America is in the process of converting Mexicans, Cubans, Guatemalans, Ecuadoreans, Argentines and many others from Latin America.

Conversion to the English language is one of the big assignments of the loyalists from the nation-state called America, and it is a task shared by many internationalists from the U-World (see below).

Countries that wish to avoid Salmon Day must make English mandatory as their second language. Here is why English is so important today. When two non-native English-speaking people get together to transact business, it is more than likely that they can only communicate in standard American English, or sometimes in British English, Australian English, Indian English, South African English, West Indies English, or one or more of the thousand English Creole variants around the world.

U-World

Such memory does not simply happen among the world's peoples. It is always motivated, always mediated, by complex political forces and economic processes, especially the United Nations and nongovernment organizations (NGOs), some 170 nation-states and 500 multinational enterprises, and free, fair and closed markets. Here are several examples:

- World federalists searched for a better world after World War II and the Cold War. They offered as the path towards enlightenment the UN, the UN Commission for Trade and Development (UNCTAD), UN Economic Commission for Latin America (ECLA) and many other world institutions.
- International bureaucrats negotiated free trade deals. They provided a new righteous behavior through the World Trade Organization (WTO), the General Agreement on Tariffs and Trade (GATT), North American Free Trade Agreement (NAFTA), the European Union (EU), the Common Market for southern South America (Mercosur) and the Asian–Pacific Economic Consensus (APEC).
- Rooted cosmopolitans invested billions of dollars in manufacturing plant and equipment, marketing and financial assets in the US, Europe, Japan and the rest of the world as the creed for a brave new world of partnerships between nation-states and multinational enterprises.

Let's call these important actors in the international economy the U-World. Many are warriors from the Cold War who, with the rest of us, find themselves on the verge of going from something to something else. They seek to lead us into an unknown future without any precise information about where we enter, how we pass through and where we come out.

To date we do not know the questions. If we do not come up with answers, America's leadership role will be squandered. Similar to the UK after World War II, the former Soviet Union and Russia after the fall of the Berlin Wall in 1989, and Mexico after the collapse of the peso in late 1994, America will suffer Salmon Day, a tragedy for its people and friends throughout the world. Avoidance of Salmon Day is America's big assignment.

U-World versus loyalists

The memories of global materialism within the U-World by federalists, bureaucrats and cosmopolitans starved out the political reality of loyalists who govern nation-states. The former did not want the latter's story to be passed on. But faced with the vivid recall of emasculated cultures, lost jobs and polluted environment the loyalists had no choice but to take their case to the public. Today, the cultural wars for public memory are between the internationalist U-world and loyalist nation-states.

Who wins, who loses? According to Gary Taylor, a Shakespeare scholar at the University of Alabama and the author of *Cultural Selection*, victory or defeat depends on a society's respect for its cultural transmitters. Pop economists and the academy are in low esteem today, and we are suspicious of the media, business executives and international civil servants. For globalization to endure, someone, some group has to say these crucial words:

> *Remember this about economic growth, free trade and global enterprise capitalism. It's our post-World War II economic story. The fundamental things go on – political risk analysis, foreign direct investment and tax breaks by governments – even when a few items change, such as the direction of trade and investment, the willingness of government to intervene in the national economy and the comparative strength of multinational enterprises.*

During the last 20 years, loyalists lost the cultural wars to the internationalists because the former surrendered the mechanisms of cultural selection to the U-Government, Global Business and Big Advertising. The latter made no secret of their contempt for political nationalists, economic protectionists and cultural purists. By discrediting the expert authority of loyalists, pop economists had to run for cover then come back by dressing up their nostrums with new images – for example, strategic trade theory. All to no avail.

What are the rememories? Strong, memorable economic artifacts have arisen in industry clusters: BMW, Toyota and Honda in the automobile industry. According to Morgan Stanley, a US investment bank, some firms have created sustainable competitive advantage worldwide; BMW through differentiation, and the Japanese firms by low-cost production.[6] All of them practice 'localisation across markets' rather than true globalization. The UK firms on this list are Glaxo Wellcome, Unilever, Reuters, Spirax Sarco and Bass. Firms from elsewhere include GE (US), AES Corp (US), Kuria Water (Japan), Cone Mills (US) and Nan Ya Plastics (Taiwan).

Virtually all of these multinational enterprises are connected to their home countries through ties of loyalty, patronage, the law and taxes owed to the sovereign. They are linked to host countries through investments, jobs and taxes due to local governments. These divided allegiances of business firms breed conflict among nation-states over jurisdiction, and the ensuing mess encourages resolution in forums set up by the U-World.

America still has not come to terms with decisions made by WTO/GATT panels and the US–Canada bilateral commission when these orders conflict with US Congressional law on international trade, economic injury and dumping. Thus the strongest multinational firms cut their own deals with foreign governments.

Win–win world

The U-World is at the end of the beginning of a new win–win world market era for global oil and natural gas enterprises. The industrial output most likely to be remembered is that produced by oil buccaneers, working in the challenging, rewarding environment of exotic locations as Kazakhstan, Peru and Indonesia. The oil majors face a big problem: 'replacing the giant fields which for decades have generated much of their profits. Their problem is not so much a scarcity of oil or gas – there are still ample reserves of both fuels. With oil, the trouble is that politics keeps Western oil firms out of the remaining giant, low-cost fields (such as those in Saudi Arabia and Iraq). With gas, the difficulty lies in finding buyers willing to sign the 20- or 25-year contracts which make it worthwhile to invest in a new field.'[7] These 'Seven Sisters' are spending a lot of time swimming upstream and some will get screwed in the end by governments. The oil and gas deals work only if pipelines and terminals are built to export oil – and that in turn depends on the turbulent politics of loyalist nation-states – in the former Soviet Union of Central Asia; near the Caspian Sea (e. g, Azerbaijan and Iran); and across the Baltic Sea (from Ukraine, Russia, Armenia and Georgia in the north to Turkey, Bulgaria and Romania in the south). Let's remember this as Salmon Day.

Mobil Oil

'Lucio Noto, the head of America's Mobil, is an oilman of the old school: a gambler and a power broker.'[8] The oil buccaneer acts like a speedboat rather than with all the agility of a supertanker. This 'straight-talking son of an Italian immigrant who professes not to take his job too seriously ... recently committed billions of Mobil's dollars to a string of deals in [the world's] exotic locations'.[9] Noto fires first in several similar loyalist nation-states all at once. Then aims with 'a 25% stake in Tengiz, a giant oil field in Kazakhstan, for around $1.1 billion; a stake in Natuna, another big gas field in Indonesia, and also one deep in the Peruvian rainforest, both for an undisclosed amount'.[10] Then Noto fires again 'looking for buyers for liquefied natural gas from Ras Laffan, a huge field in Qatar in which [Mobil] has a 30% interest'.[11]

Oil executives continually answer the key question, 'What do I get?' for persuading presidents and potentates to give the majors access to oil and gas around the world. It's chaos out there in world oil and gas markets. The 'Seven Sisters' must deal with nationalizations in Latin America, coups in sub-Saharan Africa and post-Communist troubles in Central Asia, the tiny events that lead to fundamental changes with no limited number of outcomes. Theirs is a nonlinear world. Therefore, decision making under uncertainty calls for oil executives to produce industrial output under conditions of graphic memories.

Noto's rememories are John 'Bet-a-Million' Gates, who helped start Texaco at the turn of the century, and Jean Paul Getty and Armand Hammer who wagered personal and family fortunes on oil deals. Oil executives are prone to remember risky projects associated with persuading loyalist nation-states that the latter's interests lie in allowing oil from troubled regions to flow to the West. The oil executives are the transmitters, editors and historians of the industry. They are experts who remember the past for a reason – gamblers might lose their shirts, but power brokers with luck and skill might 'Win-a-Billion'.

Other people who are not so immersed in the romance of the oil and gas industry remember these direct investments as the uninterrupted availability of fuel for cars and trucks, the price of oil at the pump and the tax revenue collected by local and national governments. When things go well these derivative benefits are powerful commitments from rooted cosmopolitans to the *status quo ante* of the U-World. When things don't go well – for example, the petrodollar crisis of the 1970s, the sovereign debt debacle of the 1980s and the lack of Iraqi oil in world markets in the 1990s – memories of citizens (and their coalition supporters from abroad) are transformed with demands for loyalist nation-states to do something about the lack of supply, higher prices and polluted air.

Intellectual assets

The problem for oil executives is not to confuse data with knowledge and information technology with information. Their difficulty is in sifting ideas, sharing information and exploiting knowledge, especially intellectual capital embedded in family connections, personal relationships, managerial teams and business network alliances. These executives want to use more of their firm's hidden values so they can respond more quickly to market pressures from all over the world. This is the essence of the intellectual revolution that now controls the destiny of the oil majors and other global firms in the quest for resources and market share across the globe.

'What needs to be done?' Find more oil for the insatiable appetite of the world's car-hungry population. 'What do we get?' Access to known oil reserves in the former Soviet Union. 'What can we do?' Build pipelines, pumping stations and deep water terminals between the Caspian Sea and the Black Sea for trans-shipment of oil to industrial Europe, Japan and the US. 'Tell me what you are going to do on Monday that's different?' Look more carefully at geological information about the flow of oil and natural gas, and make more careful decisions about rates of lifting and flow through the pipelines.

The technology revolution places new value on workers, and without human capital new technology would be worthless. Knowledge workers are more important than factory equipment and industrial equipment. Today, memory is best served by combining workers who have a global vision with new information-based technology so the search and extraction of oil can be done better and cleaner than in the past.

Many older US firms used information, not technology, to become knowledge-based companies. Once they reimaged themselves, they began to practice intellectual asset management. Here are a few examples:

Monsanto

Immodestly, Robert Shapiro, the brainbox boss of Monsanto, says 'We don't know what we're doing most of the time. We're making it up.'[12] The firm is taking genetically doctored seeds to market – such as soyabeans, potatoes and cotton to farmers – in the US, China and India. Right now these seeds are considered safe, a clean alternative to the depletion of the earth's resources, but they could be the biological equivalent of nuclear power and be banned as unfit for human consumption. Shapiro's work is to create a revolution in agricultural biotechnology and build a new industry.

Owens Corning

CEO Glen H. Hiner says 'Our real business is knowledge. We were focused on materials, but customers are asking for more than products. In the hot home-improvement market ... consumers think about projects, not products: adding a room, fixing the roof. *Stuff* is secondary ... customers for ... composites ... increasingly want component systems rather than just parts. That insight was the seed of a big idea. By leveraging the Pink Panther brand ... Owens Corning could blow away the old givens ... and offer one-stop shopping and solutions to the problems projects pose. Information is becoming more valuable than goods. Major accounts – retailers like Home Depot and Lowes – needed information more than sales pitches. They wanted sophisticated logistics ... and trained sales personnel ... knowledge-based selling. There's no stopping halfway to becoming a knowledge company: you either are or you aren't. Knowledge pulls the product, not vice versa.'[13]

US manufacturing

'What do we get?' New knowledge, better information, more sales and higher returns on assets employed. American firms back from the dead. Heroes of US manufacturing with unbeatable quality, unbelievable productivity gains and breathtaking new technology. The corporate leaders have kept American factories busy in the face of high-wage competition from Canada, Europe and Japan, and low-wage competition from Mexico, Brazil, China and the countries of South East Asia.

'What can we do?' The CEOs battle for orders with foreign companies. They get to the domestic market faster than overseas competitors. Thus they keep industry thriving in a high-wage country, such as the US.

'What are they doing that's different on Monday morning?' Examples:[14] Remmele Engineering in Minneapolis, Minnesota uses machine tools of unusually high velocity to make parts whose reliability – in medical devices and jetliners – is a matter of life and death. Deneb Robotics in Auburn Hills, Michigan pioneers simulation software that lets manufacturers try out new processes cheaply on a computer screen before they cut metal. Westt Inc. in Menlo, California take on a supplier's design work and set up its online manufacturing links.

All these American firms have gone through an intellectual revolution in which ideas are invaluable assets for present and future growth. The companies are worldbeaters. Profits are strong, jobs multiplying, product quality much improved. Many US firms have finally fielded the right formula for generating wealth and prosperity in the highly competitive global economy.

Cultural reimaging

America is strutting its stuff. Unemployment is low, inflation inconsequential, the stock market booming. The mantra: 'privatize, deregulate, no market interference, weed out firms that can't get better, downsize, capture market share at home and abroad.' It's the triumph of *laissez-faire* economics. America is triumphant!

But is the American formula the best model or must others find their own way?

Cultural remembering within the US is reimaged in three ways. First, it is mediated by American politicians (Nixon's Wage and Price Board, Ford's commitment to 'Whip Inflation Now', and Carter's efforts to tax old oil and new oil at different rates). Second, it is reconciled by American academics such as Dean George Schultz and Professor Jack Grayson who kept a lid on prices and wages for Nixon. And third, it is conciliated by American executives such as Marc Rich (who found an illegal way around Carter's old–new oil tax policy and subsequently fled to Zug, Switzerland), and James Brady who as Reagan's Secretary of Treasury created 'Brady Bonds' to back illiquid sovereign debt from the governments of the Latin American countries, especially Mexico, Brazil and Argentina.

The U-World is a sustained media event to recover the freedom of the nineteenth century when England alone mandated that foreign trade and bond investment be free, unfettered and without government control. The American Century, in which sovereignty is at bay is a dispute about US leadership in the West, French resistance to English as the universal language for commerce and diplomacy, and how to integrate Mexicans, Chinese, Russians, Moslems and others into the WTO/GATT system. Neither politics nor economics escape from the need for editors who transmit collective memories about America – namely, the first multicultural international country whose citizens recoil at giving up their loyalties to the Constitution, its Bill of Rights, the Stars and Stripes, the checks and balances of divided government, and free enterprise capitalism always at war against federal and state regulation.

The American Century is an editorial dispute about the application of a premodern eighteenth-century text to the multipolar world of the very late twentieth century. Competing texts from world federalists and international bureaucrats coexist with that from the 'Founding Fathers' at Philadelphia, and the public is free to accept or dismiss them with impunity. Supporters of a U-World want to diminish the offending constitutional texts from America, Canada and Europe. Champions of loyalist nation-states demand eradication of the offending 'one-world' theories.

Cultural memory is combat at the highest level in which intellectual analysis influences governments, armies, business firms, citizens, undocumented aliens and many others. We have no choice in remembering both the U-World and loyalist nation-states. For the balance between globalization and localization to endure, someone, some group has to say these crucial words:

> *Remember this about economic growth, free trade and global enterprise capitalism: The U-World freed us from world war and gave the West and Japan cheap Arab oil to power economic growth. If we forget that, the International Monetary Fund, the World Bank and the World Trade Organization will cease to serve the world; economic warfare among nation-states will become instead the way of the world. You must remember this too: loyalist nation-states accepted punctuated harmony among themselves and with the U-World as the price to pay for joining the bandwagon of economic growth sponsored by the Organization for Economic Co-operation and Development (OECD).*

Nevertheless, we should not continue to recollect the U-World on U-World's terms. Also we should not continue to recall loyalist nation-states on loyalist nation-states' terms. Instead, both memories are changing over time in response to external forces. Both are formed out of new objective information, data and material. They are open to subjective manipulation from the folks who are political citizens of one nation-state, but who are cultural and economic 'citizens' of many nation-states or the U-world. These successful investors maintain a dual citizenship with passports in one hand, and networks of families and friends in the other.

Rooted cosmopolitans

Let's call these multipolar 'citizens' rooted cosmopolitans. Here's what they are about:

Disseminators of the post-World War II era's view of free trade (that is, WTO's principle of most favored nations).

College-age rebels over Vietnam, then parents without secure middle-class jobs, and finally establishment figures in government and business.

Deal makers in the Asia–Pacific region as it rose to pre-eminence in international trade over the North Atlantic region.

Vidkids who are successful promoters of the computer-based digital revolution.

These experts give us collective memories, vivid rememories and reimaged remembrances. Nevertheless, their acts in transmitting memory now are viewed by the public (especially, by those whose income fails to keep up and who cannot master the computer) as the hectoring of ideologues. But the public is of two minds. If the transmitters are their successful children – that is, those who make the VCRs work, build the family's wealth, and become presidents and prime ministers of loyalist nation-states – then parents adopt as best they can home-based entertainment, E-mail: service and political mulitculturalism.

Today, in 1997 the world is changing once again. Will the U-World continue to swim in the seas and rivers of the international community and take an increasing share of the food chain, or will it swim upstream to its death? Will loyalist nation-states continue to swim along in the same oceans of the international community and accept a diminishing share of the resources; will they fight back with new strategies and tactics; or will they too swim upstream to an inevitable end.

Coming to America

What is the future for the land of Washington and Lincoln, the world's first multicultural international community, the hope of many, the scourge of some – America, the goddess of prosperity for the masses?

According to the former Librarian of Congress Daniel J. Boorston, America flourishes on the verge – as a place of encounter between something and something else – of settlement, geography, culture, technology, or history.[15] Today, America is on the verge between its decades long manipulation of the U-World for its own international purposes and the desire of its weakly governed, brawling, factious society for a world of loyalist nation-states in which the US has political, economic and cultural rights and few responsibilities. As Michael Cohen, an editor at *Dissent* says, Americans among the media elite, bureaucrats and world federalists are rooted cosmopolitans – people with several loyalties, standing 'in many circles, but with common ground' in the form of a home base, the United States of America.[16]

In the late twentieth century, America is 'a place where all the major cultures, religions and economies meet, conquering the world as it conquers us'. America is again on the verge between something and something else, in its ordeal of developing linkages between the U-World and the loyalist nation-state world, in its quest to absorb Russia and other places into the postindustrialized West, in its desire to avoid collapsing on itself.

Here's America at its best. Its transmitters of memory engage in foresight and alter its future by the choices they make. These ideologues show the rest of us how to swim among the currents of the U-World and loyalist nation-states. They offer us choice.

Here's America at its worst. Its editors of rememories fail to pass on the stories and starve our reality. Only vivid recall of past successes and failures offer us choices for reshaping the future.

America as loyalist nation-state

The business guru Kenichi Ohmae defines the nation-state as a emotional place in which government subsidizes uncompetitive enterprises, protects weakened workers and doles out aid to failing cities. Through the political process and with the ability to tax the rich in favor of the poor, Indianapolis (the capitol of Indiana) sends surplus state and federal funds to the former workers of the R.R. Donnelley & Sons plant in Crawfordsville and to the shuttered steel mills in Gary. Downtown Indianapolis is the place where the state does its business, the Colts (formerly of Baltimore) play NFL football and conventioneers embrace global materialism, not unlike Rosemont, Illinois (near O'Hare airport), the center city of Singapore and many other retail upstart urban enclaves around the world.

Distinctions blur. What remains are cinnamon donuts at Mr Donut, pan and thick style pizzas at Shakey's Pizza, quarter-pounders (or double burgers) at McDonald's, cappuccino at Starbucks, Hollywood movies, CNN and Gucci shoes in the midst of Bloomington (Indiana University),

Lafayette (Purdue University) and South Bend (University of Notre Dame). Localisms are hard to find. These high-income, intellectually oriented service centers grow increasingly similar with Oxford, England, Queens Park in Toronto and the Kanda neighborhood of Tokyo.

The work of their 'citizens' is computer-driven in a knowledge-based economy; they share the common interests of highly educated friends all over the world. These transnationals are the transmitters of memories about the U-World and over time they edit out their tenuous attachments to loyalist nation-states. Rooted cosmopolitans (that is, the middle and upper classes) see America's social security and income taxes, welfare programs and health care services as obstructions to success in a global economy. Today, they want an end to the intrusion of public life pathologies (the tedious two-year campaign for the American presidency, the antics of the Windsor Royals before the solemn civics lesson of the Queen's Throne Speech at Westminster, and the money payments from Japanese business executives to LDP political leaders), and a cessation of transfer payments to the less fortunate, uneducated poor.

Rooted cosmopolitans want to reinvent America as the home for the U-World in which local interstate highways and airports bring in the best and the brightest from all over the world, at least for part of the year, and give residents speedy travel to work and play within North America, across the oceans to East Asia and Europe, and down south to Latin America, southern Africa, the Persian (Arabian) Gulf and South Asia. Here is America at its best with a pluralist society, decentralized government and citizens free to come and go as they please.

Loyalists want to reinvent America as the home for new emotional attachments. Individual freedoms as enshrined in the Bill of Rights become group rights of the Rainbow Coalition (They take a leaf from Canada's Charter of Rights). The Stars and Stripes flies along side of the flags of African-Americans, homosexuals and lesbians, and other groups who want an end to their second-best past and a bright new future from those who have power, position and prosperity in the larger American scene. Free enterprise capitalism must give something back to the poor, the dispossessed, the disenfranchised. Here is America at its worst with a continual fight over the meaning of patriotism (is there any left?), and whether the center should be weakened or strengthened now that the Cold War is indeed over.

America today means the United States of America as we know it has had its day. The loyalists bet they can reform the country, give its older symbols new meaning and maintain themselves as vocal allies of a deconstructed past, an emerging present and an unknown future. They want to shake the money tree of business until wealthy executives squeal with both pleasure and pain. Loyalists believe they are on a mission (from God?) to create a new American civic religion for a new American people who will continue to run the world in America's second century. Theirs is loyalty to America as the world's most important nation-state.

America today also means the US must find a new role to play in the U-World. Rooted cosmopolitans bet they can have their cake and eat it too. They insist on their rights as world 'citizens,' but forego their political, civic and religious responsibilities. They too believe they are on a mission (from someone to someone else?) to create a new civic order for the people who live in the Asia–Pacific region, across North America and well into the center of Europe. Theirs is loyalty to an extra-legal America as the world's most important civilization since the decline and fall of the Roman (and Byzantine) Empire so many centuries ago.

Which vision of America will swim the ocean's currents without climbing the rivers to inevitable death? Does Salmon Day come to all civilizations? Or are there ways of escaping fate?

One-time events

The U-World forced three one-time events on America. These are as follows:

First, the need to remain the dominant power in the West and indeed across the globe forced the Reagan administration to finance America's military buildup in the 1980s through deficit financing. Taxes were cut but spending went up. Some 16 years later America still lives with the bill to service the debt even though the Soviet communist threat to America's security is gone. Worse still, America lives with the memory of being an imperial power, and acts like one in the United Nations, within NATO and in its dealings with other sovereign nations. The U-World has been unable to stop American loyalists from dictating to the world.

Second, the need to keep Japan on America's side in the Cold War and the need to let other East Asian nations grow economically exposed the American nation to their imports, far beyond the country's capacity to pay for them with exports. This deficit in the merchandise account of the balance of payments cheapened the US dollar and made investments in the US cheap. Foreigners, especially the Japanese, started buying up American properties with little thought to how they would grow the business in the US. In the long run, this was a mistake. When the US dollar started rising in value, foreigners lost money as they sought to get out of their investments. Notwithstanding all these changes in the relative value of the US dollar in terms of the Japanese yen, American loyalists have been unable to get the U-World to slow down aggressive conquests of market share in US markets.

Third, the need to become competitive against Japan, Koreans and others in East Asia came as a surprise to Americans. First, we set up artificial barriers against their products. When these did not work, we sought an accommodation with our friends and allies – that is, limit the flow of foreign imports, reduce barriers overseas to US exports, pay more for US military support, and fight dumping and economic injury before the US International Trade Commission (ITC) and at the WTO/GATT. None of these actions resolved the problem and America continues to suffer the ups and downs in its trade deficit. American loyalists looked for guidance from foreigners, adopted some of their managerial reforms and foresaw a better day tomorrow.

Optimists, such as Michael Moynihan, argue that these one-time events occurred in the past, they made the US stronger and it will out-compete other nations in the next century.[17] Doomsday experts (such as Clyde Prestowitz, Lester Thurow and Paul Kennedy) say it is not so as a unified Europe or capitalist development states in East Asia wrest dominance of the world economy from the US.[18]

Let us be careful with these extremes. The more likely outcome is as follows: America is on the verge of going from something to something else. America will go from being the predominant center of the industrial world to the strongest, but not the overwhelmingly dominant economic power among East Asia, North America and Europe. The center will hold as some power is passed to peripheral areas, east and west, north and south, and across the globe. And Salmon Day will be avoided. This is the big assignment for loyalists and one that might even capture the imagination of the U-World, too.

America's big assignment

America's urge to carry out big assignments is likely to continue because, once ingrained, it tends to propagate itself both at home and abroad. Our collective memory of America is a land of big egos, big malls, big midriffs, big cars, big roads, big garages, big houses, big refrigerators, big ice cream containers, big athletes, big hats, big shoes, big spending, big people. Together, we have the following remembrances: America is a land with the big ideas that are tempered with diversity and choice.[19] America is the only nation in the world to take on the really big assignments.

Remember this about the big idea. America is on the verge of something big to something bigger. Americans speak English and the English language is now the operating standard for global communications. Already 80% of the information stored in the world's computers is in English and 70% of the material on the Internet is in English.[20] America will choose as its Asian, European and Latin America free trade soulmates in the new world of industrial power those countries that use English as their means of communication. This is already true in many western European and East Asian countries, and some Latin American countries. The crucial holdout remains France, whose president is frightened about what the Internet may do his country's language. Nevertheless, the decisive factor in postmodern history is that Americans speak English.

Costs and benefits

How should America weigh the costs of attempting to create a pan Asian–North American–European community against the benefits of less political and economic conflict in the future?

This is a case of decision-making under uncertainty. If all the countries involved join in the effort in a spare-no-expense way (by yielding sovereignty over all economic matters), the possibility of further conflict would be greatly minimized. If the countries involved join in the effort on a slower basis, a greater possibility of political conflict would remain. How much greater is unknown.

Decision theory cannot create certainty where none exists, but it offers a way to organize one's thoughts and gather the information needed to make better decisions. Consider free trade. The best policy is for the US government to use NAFTA to link up with all countries that are interested in free trade with America.

Remember America has choices. One option is to give birth to a North Atlantic Free Trade Area. America and the other NAFTA countries could create a larger free trade area by joining up with the EU, the European Free Trade Area (EFTA), and key countries in eastern Europe. If America chooses Europe first, it does so because a common civilization exists between California and the Polish border areas with Germany. European immigrants to the US who are still the majority of the American population play their common kith and kin, language and religion cards.

A second option is to give birth to an Asian–Pacific free trade area. The NAFTA countries could create a larger free trade area by extending NAFTA benefits to South Korea and Singapore, and deepening their trade and investment ties to all Asian members of APEC. If America chooses East Asia first, it does so because a future civilization is being created between Bangkok and New York City. Asian immigrants to the US who are still making their way through the corridors of power play their wealth, industry and education cards.

A third option is extend NAFTA to Mercosur, the common market for southern South America, and to the other countries of Latin America. If America chooses Latin America first, it does so because pressures from its Mexican and Latin American citizens are too great to ignore, and they are intent on creating a joint Anglo-Latin commonwealth throughout the Americas.

Option one has more political, economic and cultural benefits today *vis à vis* option two. America's dominant military position on the European mainland strengthens its traditional ties with the peoples between the North Atlantic Ocean and Vistula River in Poland's 'Asian' lands. Also America gains insider status in the discussions about how Europe adds new members to the latter's highly successful European Union.

Option two has more longer term political and economic benefits *vis à vis* option one. America's dominant military position in the Pacific region remains to back up allies and friends during communal strife over market opening measures, oil under the ocean and human rights problems. Also America gains new strength from the emergence of a Chinese entrepreneurial class who bestride both sides of the Pacific Ocean.

Option three has one long-term political benefit: it keeps the Mexican border areas calm. America gains little in terms of economic benefits because the Latin American countries are far behind Europe and East Asia in terms of industrial output, productivity and economic growth. However, America does affirm the Mexican march towards US citizenship and the quest for other Latin Americans to participate in the American dream. But we incur a cost. For a long time in the future, about 20% of the population in America's heartland will speak Spanish rather than English as their primary language of communication.

Decision theory offers us a way to reflect on our choices and make the best decisions we can with the information we have in front of us. Decision theory offers us a way to organize ourselves to beat the fate of Salmon Day.

Salmon Day

Salmon Day is the experience of spending an entire day swimming upstream only to get screwed in the end. Mobil Oil, Monsanto and Owens Corning fought the global fight, gained market share and won support on Wall Street. Smaller US manufacturers took their newest technologies, applied them in a high-wage country and still made money. None of these firms were victims of Salmon Day, and they represent American successes in the late 1990s.

Here are some summary thoughts. First, the transmitters of America's common knowledge about the big idea called America offer the following: the global, largely American life style; English as the universal language of communication; and America's newest big assignment – namely, the creation of a free trade area within East Asia, North America and Europe.

Second, the editors of America's common knowledge are in the process of reimaging their intellectual thoughts, conclusions and policy prescriptions concerning the U-World and world of loyalist nation-states. Some of their vivid memories will remain while others will be remembered as one-time events whose costs have been fully discounted and whose benefits have spread throughout American society.

Third, managers and others in the professional class must decide whether America expands free trade and with whom it joins in completing its big assignment for the twenty-first century. To decide for Europe or Asia means America continues to swim the oceans without serious risk to its fate. To decide for Mexico and Latin America means America faces a dangerous swim upstream and the possibility of getting screwed in the end.

All agree that America is the only country that can be both a political nationalizer and economic globalizer. It can use its nationalism to strengthen the global, largely American life style in Europe, Japan, China and East Asia. This makes the US unique among the nation-states of the world. Mobil Oil, Monsanto and Owens Corning are the true beneficiaries of the American market-rationale approach towards the international economy. Their worldwide strategies may be signposts to the future for American, European, Japanese and overseas Chinese firms.

Wisdom: decision theory lays out possible choices. Pay attention: one choice leads to Salmon Day, and the others lead to a second American century. Open the doors: America makes a choice to defend its most important interests.

What do I get?

Here is our to do list:

- Buy into the big assignment.
- Accept loyalist nation-states as crucial to the U-World.
- Build on the memories of global business and big advertising.
- Reimage ideas within the global, largely American life style.

- Recognize rooted cosmopolitans as 21st century citizens.
- Create a pan Asia, North America, Europe stream of business.
- Use decision theory to weigh costs and benefits of options.

Remember these words: Salmon Day is not inevitable for America.

Chapter 2

Nafta and Mercosur

'America did not get any significant increase in domestic jobs nor did it lose any significant jobs because of NAFTA.'[1]

Raul Hinojosa Ojeda, research director, North American Integration Development Center, UCLA.

'Mercosur built an artificial world of economic growth on the basis of inefficient industries prospering behind a wall of trade barriers.'[2]

Alexander J. Yeats, principal economist, International Trade Division, World Bank.

Societal memory

Memory mothers Crown Corporations, babies foreign direct investments, and nurses the continentalism of peoples along the border between the US and Canada. Also memory soothes the minority French 'nation' along the St Lawrence river of Quebec and reassures the majority English (immigrants from Britain, Ireland, Europe and South Asia) of the rest of Canada. Both Canadian solitudes live within 100 miles of the border with the US. English-speaking Canadians, especially, are vulnerable to being overwhelmed by American TV, the mass market culture of bigness, and the demands for increases in market share from the largest firms in the world. Unfortunately, memory sustains American forgetfulness about real differences in Canadian politics, and apparent similarities in cross-border, North American socio-economic truths. Therefore, memory demands global Canadian executives be familiar with American cultural icons whereas their counterparts in the US can overlook Canadian cultural institutions.

What do all Canadians remember?

- Standard American English in business communication;
- *The Wall Street Journal, Business Week, Time;*
- CNN, PBS, ABC, NBC, CBS;
- Seinfeld, Friends, Frasier, Roseanne.

And, of course, Burger King, Pepsi, Disneyworld and the Internet.

What do English-speaking Canadians want to remember?

- Ties to the English crown, parliament at Westminster, Oxford and Cambridge.
- *The Globe and Mail, Macleans, Saturday Night.*
- Canadian Broadcasting Corporation (CBC).

And, of course, Lorne Greene, William Shatner, Michael J. Fox, John Candy – all Canadian actors who made it big in Hollywood. Margaret Atwood, North America's most important novelist who lives and works in Toronto. And many others in broadcast journalism, country and western music and other Canadian cultural institutions that extend into the US, dominate memory in both countries and come back into Canada as a domineering but enjoyable global, largely American life style.

The big idea of America goes into Canadian homes through TV, and gets partially repulsed by United Empire Loyalists (They are Canadians whose ancestors were Americans before the American Revolution of 1789 and who were shipped to Halifax and York, now Toronto) who fight the good fight in trying to slow down inexorable changes in Canada.

Neither the new Maple Leaf flag, the new national anthem, *O Canada,* or the new Constitution with its Charter of Rights are able to stop the slow decline of the Canadian difference among English-speaking Canadians. Of course, French Canadians have their own language to protect them from America's big idea, but even they must speak English to do business with other North Americans, most people in Asia and many in Europe.

Nevertheless, both Anglophone and Francophone Canadians eat Yankee food – chips and salsa – and prepare a beaver tail (fried dough covered in maple syrup) to watch on TV the Super Bowl, America's NFL football championship, and the Grey Cup, Canada's CFL football classic.[3] The games, the ritual of watching them on TV, and the eating of snack foods while watching the football games on TV are icons of the global, largely American life style. Moreover, 'a growing minority of Mexicans now watch American ... football on TV, and when they do they tend to prefer Yankee snacks – pizza, hamburgers and potato chips chased with Cokes'.[4] In this regard, Canadians and Mexicans are just like Americans.

Immigrant memory

Memory mothers xenophobia, babies nationalism, and nurses the misunderstandings of peoples along *La Frontera* (the raw border lands between Mexico and the US). Also memory sustains undocumented Mexicans who see *El Norte* as their hope for future. They travel from the impoverished Indian lands in the south, through the Mestizo lands in central Mexico to the frontier waiting for the cover of night to cross over into the US. In the end, they leave the land of *tú*, but they keep part of their past with them on their sojourns in the US.

The long-settled Mexicans in the US spend 10–11 months in their adopted country and bring their cash home to convert their 'adobe huts ... into trim cement-and-tile houses, many crowned with satellite-dish antennas'.[5] They are home (in Tehuixtla, a village in the state of Puebla, about 140 miles south-east of Mexico City) for Christmas sitting on their porches shaded with blazing bougainvillea, and eating home-made tortillas, refried beans, guacamole with their family.

These Mexican citizens and resident aliens of the US wonder whether, after 30 years of going back and forth, now is the time to become US citizens. By reasons of birth their children are already American citizens. If they did not own their homes in the village, they too might be tempted to take the loyalty oath to *El Norte*, a pact with the gringo who took one-half of Mexico in 1848 and who grabs for the soul of Mexico today.

At no time during their Christmas vacation in Mexico do these resident aliens look forward to the long trip back to their work in the colder climates of New Jersey, New York or Chicago. Unfortunately, no work exists for them in these small meso-American villages of Mexico. They must go. Mexicans must face *la migra*, the immigration gauntlet of the US, before they can cross *La Frontera*.

With them comes their cultural capital (their differences in social choices, economic efficiency and political stability).[6] Do they marry? Do they work? Do they save? Do they buy homes? Do they educate their young? Even though immigrants share the hardship of the journey from the land of *tú* to the land of you, once in the US they look for ways to begin creating wealth, prosperity and a new way of life. Theirs is a formidable achievement, and one that accelerates the globalization of tastes among Americans in the north and Mexicans in the south.

Undocumented workers, resident aliens and new citizens – immigrants all – buy into the big idea of America, its chaotic political life in the heartland, its deep socio-economic ties with Canada and Mexico, and its wider commercial links to the nations of South America, East Asia and Europe.

Historical memory

What do Mexicans remember?

- The loss of California and Texas to the US.
- The loss of workers to US farms, factories, restaurants.
- The loss of economic independence to a US-dominated NAFTA.
- The replacement of Lola Beltrán, an icon of traditional music, by Madonna.

And, of course, contemporary American culture: McDonald's, Coke, Disneyworld, Bart Simspon, NFL football and the Internet.

The big idea of America goes into the homes of affluent Mexicans through HBO-*Olé* and CNN. However, it becomes a smaller, finer calibrated idea when Mexicans remind themselves who they really are: inheritors of ancient Indian cultures mixed with proud conquerors from Spain in a new meso-American race, called Mexican. Both the Spanish and many Indian languages protect loyalist Mexicans as nothing can protect English-speaking Canadians from the big reach of America across the borders of North America.

Dual citizenship

Upper-class Canadians, rich Mexicans, titled Europeans, high Chinese party officials and other elites with position and power view rampant consumerism as an expression of egalitarianism, an imported alien value from unrestricted globalization and the penetration of American culture worldwide. Satellite television lets everyone (rich and poor alike) see goods, life styles and ideas about the abundant life. The elites fear they are losing control to foreigners. The global, largely American life style has become a cultural stranger to them. They resent the talk of their leaders about growing homogeneity as if the well-watered prairies of Canada are the same as the water-starved high plains of western Kansas and eastern Colorado, and the latter are the same as the semi-desert of central Mexico. In Calgary, Denver and Guanajuato the super-rich are seceding from their cities, provinces or states and nations, and they are building enclaves of super-privilege and super-protection. They may hold national passports, but their loyalties are to their networks of family connections and personal relationships within North America and Mexico, and across the world. These dual citizens benefit from technology and globalization while at the same time they complain about the impact of alien ideas on the local culture.

Win–win–lose world

The U-World is at the beginning of a win–win–lose NAFTA era for telecommunications firms. The industrial failure most likely to be remembered is that produced by Iusacell, the Mexican acquisition of Bell Atlantic (BA), which led Telcel, an arm of telephone giant Teléfonos de México (or Telmex), to get way ahead in the cellphone business. 'BA's troubles are a cautionary tale for multinational companies that thought investing in Mexico would be a snap under the North American Free Trade Agreement. BA stumbled over what is still a vast gulf in business cultures. And it was slow to recognize the shift in Mexico from clubby government-business relations, which guaranteed success for anointed deals, to a system constrained by public scrutiny.'[7] The BA-Iusacell parent–partner relationship was plagued by a clash of corporate cultures, and by brash American executives who were unable to work with more subtle Mexican executives.

Lastly, Iusacell's former Mexico owner, Carlos Peralta, was tied to the difficulties of the previous Mexican president, Carlos Salinas de Gortari, and Bell Atlantic found itself unable to overcome this political setback quickly. Let's remember this last problem as Salmon Day.

Telmex

Many wired, fixed wireless and cell telephone firms in Mexico face Salmon Day because of the peso devaluation and the continuing Mexican economic crisis. Most Mexicans simply cannot afford traditional installation costs and are not eligible for extended payments through credit cards. Instead, Telmex offered them a prepaid cellular phone service with no monthly service charges. Customers pay a high charge per call, but Telmex has no back office costs and no past-due accounts. So the firm makes money on this better idea.

The problem for Telmex and other large Mexican firms is where to put its investment dollars. So many are tied to central Mexico and Mexico City. However, the 130-mile wide strip that stretches 2100 miles from the Pacific Ocean to the Gulf of Mexico – known as *La Frontera* or The Border between the US and Mexico – 'may be North America's fastest-growing region. With 11 million people and $150 billion in output [half of Mexico's GDP], it is an economy larger than that of Poland and close to the size of Thailand. Tijuana [is] the world's TV manufacturing capital, churning out 14 million sets a year. With total wage, rent, and electricity costs running as low as a quarter of those in the US, the border became the cross-road for global bargain hunters',[8] such as Lucent, Sony, Daewoo and GM. In fact, the Delphi division of General Motors has an R&D center in Ciudad Juarez (across from El Paso, Texas), which serves customers all over the world.

These border plants (or *las maquiladoras*) operate in the dollar economy, and they are doing well even with the peso devaluation and continuing economic crisis in Mexico. But their wage payments add little of value to Mexico's real economy. Those Mexican factories that are further south in central Mexico operate in the peso economy, and they are not doing well.

They are being squeezed by high interest rates, heavy taxes, stifling regulations and an inability to enforce contracts and collect debts. Some have become sub*maquiladoras*, or local firms that support the assembly plants in the north. The essence of a forthcoming intellectual revolution in Mexico is to give up the peso and dollarize the entire economy. Otherwise, everything south of The Border will suffer Salmon Day.

'What needs to be done?' Sell more cell-phone services in northern Mexico. 'What do we get?' Access to local and US customers. 'What can we do?' Build fixed wireless to local executives and sell prepaid wireless services to employees. 'Tell me what you are going to do on Monday that's different?' Look more carefully at market segments and target them with cost-effective telephone products. Today, memory is best served by combining Mexican desire for better telecommunications with existing and new technology so The Border indeed becomes the most important 'tiger' in the developing world.

Panamco

Pan-American Beverages Inc. (or Panamco) purchased the Venezuelan bottling venture between Coca Cola and the Cisneros family, and it became the second-largest Coke bottler in the world. Panamco already has a dominant market share in Mexico, Brazil, Colombia and Costa Rica, and Coke considers it an 'anchor' bottler with deep local ties, large capital budgets and a finely tuned distribution system. The acquisition makes the bottler a $3 billion (in US dollars) company ready to extend its bottling expertise into water, beer and other drinks.

Some Mexican firms have gone through an intellectual revolution in which free trade ideas, such as NAFTA, are invaluable assets for present and future growth. However, none of the Mexican companies are worldbeaters as is Panamco. Many Mexican firms are finding it extremely difficult to move from the peso to the dollar economy and still survive as competition heats up with American, European and Japanese firms. Their alliances with US firms fail because of very great cultural differences between the US and Mexico.

Mexico and the big idea

NAFTA encouraged *norteños* (or Mexicans who live in the north of Mexico, especially along the border with the US) to prefer North American over traditional Mexican life styles. Nowhere is this more evident than in how the Christmas holiday is celebrated within Mexico. Traditional Mexico prefers the Catholic Christian story of the baby in the manger on December 25 with gifts being given on Three Kings Day (or January 6). Both Canada and the US, and now North American Mexicans, prefer the contemporary concept of major shopping trips to Wal-Mart, the artificial Christmas tree in the house, the Santa Claus story for children and gifts given on December 25.

Some of this contemporary Christmas culture has seeped as far south as the Zócolo, the main square of Mexico City and the seat of power and prestige within Mexico, where northern European and North American images of Christmas dominate public space. This display of lights and decorations on Mexico's most important government buildings is equal to how Americans create their gigantic holiday displays in privately owned shopping centers.

NAFTA encouraged Mexico to switch its allegiance from Spanish-speaking Latin America to the global, largely American life style of the US. No longer would Mexico suffer along with Guatemala, Colombia, Peru and the others as third-world, impoverished nation-states. No longer would Mexico endure state intervention in economy along with Venezuela, Brazil and Argentina in which many crucial industrial activities (for example, food warehouses, telecommunications, steel mills, ports, railroads, etc.) were controlled and managed by government-owned firms. By the magic of the market, Mexico would end over 500 years of state enterprise capitalism, American direct investment would pour in and jobs would be created in Mexico for Mexicans.

Nothing worked out as planned. Mexico did get investment, but it was in the form of hot money in the financial markets which, when things got bad, fled Mexico. Mexico imported all the luxury and consumer goods its people wanted without exporting enough Mexican-made goods to pay the bill. Mexico's trade deficit with the US soared. Cash reserves dwindled. And the peso finally had to be devalued in the worst possible circumstances for the government, foreign investors and citizens.

Suddenly, on December 20 1994, Mexico was again a third-world, impoverished Latin American country. Its very rich put their assets in Houston or San Antonio, Texas, USA. Its middle class became substantially poorer. And its poor gave up on Mexico and came to the US. That's how the big idea worked out for Mexico.

NAFTA in the U-World

Back in January 1994 three countries in North America made a choice for the U-World. After three decades of free trade in automobiles, and after five years of free trade in most industrial goods and some investment services between the US and Canada, both countries embraced Mexico in a comprehensive free pact in goods and services for all of North America. The three countries had similar and different views of what they were getting into:

- The US wanted open markets free of local government tariffs and restrictions for the export of goods, services and capital. Also it wanted economic stability on its southern frontier with jobs in Mexico rather than jobs in the US for Mexicans.
- Canada wanted open markets, too, except in English-language cultural services, such as newspapers, magazines, radio and TV, and satellite and telecommunication transmissions.
- Mexico wanted export markets, substantial new direct investments, and an overhaul of its domestic economy.

At first, all three countries got some of the things they wanted, but they were disappointed in other things. For example, once Mexico devalued the peso in late December 1994, the local market for US goods dried up. Mexicans were poorer; the middle class was decimated; and few people had money to buy expensive imported goods from the US. Mexico went from the mirage of stability to its historical reality of booms and busts.

More Mexicans sought refuge and jobs in the US than ever before. The U-World's great success story was turned upside down as the US had to bear additional burdens in the cause of stabilizing Mexico.

NAFTA and social cohesion

To date Mexico's problems within NAFTA have had no effect on the social cohesion of the US. In fact, the opposite seems to be occurring. For example, greater Los Angeles has an emerging Latino middle class (that is, mostly Mexicans, Central Americans and others from Latin America) with households that earn more than $35,000 a year (in US dollars), or own their own homes, or both. According to Gregory Rodriguez, a research fellow at Pepperdine University, in southern California one-half of the American-born Latinos was middle class and one-third of the foreign-born Latinos was middle class by 1990. Latinos reach the middle class 'by setting up small businesses, working in blue-collar jobs and, above all, by pooling their resources with extended families living under the same roof. In all, Latinos are much more like early twentieth century Italians than yesterday's Jews or today's Asians. The newcomers are rapidly being assimilated into mainstream America. [perhaps] without losing their ancestral identity.'[9]

NAFTA and employment

America did not get any significant increase in domestic jobs nor did it lose any significant jobs because of NAFTA. According to UCLA's North American Integration Development Center, 'The impact on trade-related employment during the first three years after NAFTA is estimated to be, at the very least, a near zero net impact, and more likely, a moderately positive number ... just 2990 jobs'.[10] Raul Hinojosa Ojeda, the Center's research director, says that 'unpublished data clearly established that jobs that support exports from the United States to Mexico paid better than those that were shifted to Mexico. And the long term trend is that the United States will gain more jobs than it loses from the agreement ... [that is, with] the development of policies that can achieve sustainable growth in Mexico'.[11] Mexico within NAFTA confirmed for US loyalists what they had always believed about Mexico – the country to the south, the land of the night, *fiestas* and *tú* was not a sturdy leg upon which to build an economic partnership. Moreover, Mexican failures stalled the admission of other Latin American countries into NAFTA. In short, loyalists bought into the argument made by the Anglo-French billionaire, Sir James Goldsmith, that free trade pitted the high-income countries against the billions of impoverished workers.[12] Does global liberalization, such as NAFTA, impoverish, if not everyone in rich countries, at least the relatively disadvantaged? Goldsmith says yes.

Comparative advantage and the loyalist world

Robert Lawrence, professor of international trade and investment at Harvard University, says no.[13] He cites the theory of comparative advantage. First, Mexican wages are low because its average productivity is low. When productivity catches up to the US, so will its wages. Mexico exports goods it can make relatively more cheaply than the US – those where the relevant technology can be imported easily and where unskilled labor is used more intensively.

Second, changes in trade between Mexico and the US do not explain any of the difference between the growth rates of real output and of real compensation per worker in the US. Instead of labor giving way to capital, labor and business owners continued to get their shares of the value added of output. Unfortunately, the cost of living (especially in the price of housing) rose faster than the prices for industrial output and investment goods.

Third, NAFTA trade has not, in general, increased the inequality between more and less skilled workers in the US. Professor Lawrence makes the following points.

1. Under the quantitative impact of trade, imports and exports contain both labor and capital, and these can be viewed as additions to or subtractions from the supplies of goods and services at home. He concludes that no more than 10 per cent of the rising differential

between high-school and college workers in the US can be attributed to the quantitative effects of changes in trade.

2. Under changes in the relative prices of exports and imports, the terms of trade improvement in the prices of imports and exports did not lower the relative wages of unskilled workers in the US. He concludes that the relative demand for college-educated vs. high-school educated workers in the US only amounted to about a third of the impact of "deindustrialization" on high-impact older urban areas within the US. He also concludes that the impact of new technology and organization increased the relative demand for skilled, especially college-educated labor.
3. Therefore, the relative demand for high-school educated labor will fall fast, but only depress wages in the US by a small amount. On the other hand, the relative demand for skilled, college-educated labor will rise fast and if productivity rises fast too (which it has not so far) then global liberalization through NAFTA and other free trade pacts will be a success.

Sophisticated economic analysis in terms of the impact of NAFTA on international trade between the US and Mexico, and labor markets within the US suggest the U-World has won the day over the world of loyalist nation-states. Nevertheless, free trade is still only a brilliant economic concept that has to be translated into real-world economic policy by national governments buffeted on all sides by the desire of citizens to blame foreigners for all ills.

Unilateral departures from free trade

Does the US have the sovereign right to impose significant departures from free trade? 'Yes' says the U-World, for reasons of national defense. 'Yes' says the loyalist world, for reasons of national defense, social stability, jobs and infant industry protection, income distribution, wealth maximization, market dislocations and national advantage.

Strategic trade theorists who are within the loyalist world stress the need for the US government to protect certain crucial industries by employment compensation, export subsidies and higher tariffs. These loyalists and their U-World counterparts agree on one point: free trade is a policy that is not always correct. Under certain conditions, the case for free trade is more robust whereas at other times it will have difficulty passing from theory into practice.

In the years leading up to NAFTA, the collective American memory about free trade included the following successes.

- Twenty-five years of successful free trade in autos and auto parts with Canada.
- Almost five years of successful free trade in most industrial and consumer goods, and investment capital with Canada.
- About 30 years of successful free trade and common market integration among the nations of western Europe.

Today, the collective American memory includes the rush of US goods and flight capital into Mexico, the peso's devaluation, the American financial bailout of Mexico, the economic depression in Mexico, and the flight of unemployed undocumented Mexicans to the US. Reality has overtaken theory. Thus 1997 is the year Mexico must prove it is a worthy economic partner of the US, and 1998 is the year Mexico must grow its economy to absorb more US exports. Otherwise, Mexico faces the loyalist tide in both the US and Mexico against NAFTA.

Forecasting the future

Americans are struggling to find a balance between NAFTA and neighborhood – between their desire to preserve a sense of national identity and control over their own lives in a world where they can only survive economically if they link up to distant, soulless institutions and markets – from NAFTA to the EU to APEC – which don't reflect any identity at all. Professor Michael Sandel of Harvard University puts it this way: The US will not be able to sustain itself until it finds ways of contending with the global economy, while also giving expression to the American people's distinctive identity.[14] NAFTA is no substitute for jobs, a home, and community in a neighborhood called Lincoln Park, Pilsen, Austin, Uptown, Rogers Park and others in Chicago. Local people don't want to be swamped by global institutions. They want the political power to control NAFTA.

Perhaps local people are smarter than their politicians. In late 1996, 'U.S. exports to Mexico are starting to pick up as its economy revives. But a fundamental shift away from the U.S. in some Mexican import patterns could be longer-lasting. Mexico has been buying more electronic parts, chemicals, and other industrial materials from Europe and Asia. That flies in the face of U.S. hopes that NAFTA would turn Mexico into a U.S. industrial preserve, protected from European and Asian rivals. Economic reforms in Mexico have prompted manufacturers to look further afield for cheaper prices and better-quality goods.'[15]

Local people and loyalists agree that this is an unintended consequence of NAFTA, and one they will use to stall other attempts at free trade by rooted cosmopolitans of the U-World. The future is uncertain for many of the free trade initiatives started in the 1990s. America is on the verge once again of going from something to something else.

The saga of Latin America

Americans are of two minds about Latin America. The first set of beliefs gave us NAFTA and wants us to think that free trade from the Arctic Circle to the Tierra del Fuego is possible. It follows then that all peoples in the Americas are new peoples who endured the hardship of the journey and struggled to prosper in their new homeland. From Europeans, Africans and Asians they were magically turned into Canadians and Americans – North Americans – and Mexicans, Ecuadoreans, Argentines, Brazilians and others – Latin Americans.

The second set of beliefs wants us to forget the notion that we are all the same people. They say North Americans are English-speaking, mostly Protestant in religion and capitalist in entrepreneurial spirit. On the other hand, they say Latin Americans are Spanish- and Portuguese-speaking, mostly Catholic in religion and state interventionists in commercial activities. The two are as different as day and night, or you and *tú*.

Somewhere in between are the descendants of the Mixtecs, Aztecs, Maya and Incas, the impassive, copper-colored faces of the women and men who surround them. Impenetrable: their history has been written by the Catholicism of others from Spain, and their present is being prepared by the evangelical fervor of another group of others from North America. In Mexico, Guatemala, Colombia, Ecuador, Peru, Bolivia and Chile the mountains are the land of the indigenous populations. Immense, mysterious, poverty-ridden, ancient and closed: oil and gas seep from under ground across jungles and lunar landscapes, and product flows above ground to refineries in the great cities of North and South America.

The daily routine of work for the indigenous leaves them fatigued and confused. It is the same look they have after three days and nights of eating, drinking, dancing, fighting, praying, not sleeping during the *fiesta* for their patron saint. Reality and unreality merge into one continuous encounter with the military of local loyalist nation-states and private police guards of the oil and mining companies from the U-World. Since the indigenous secretly placate their ancient gods with human sacrifice, they think nothing of the loss of life in building pipelines, constructing roads and heliports, and building the new cities on the coast.

Deep in the background of New Mexico USA, old Mexico and all the mountainous countries of Latin America is a dreadful unity to life and death. North Americans are fascinated, repelled and simply overwhelmed by the sacrifices required from the indigenous to construct the new societies envisioned by the Spanish- and English-speaking people of Latin and North America. The idolatrous cults of the indigenous and the willingness of others to use government to their own advantage are the foundation stones for the loyalist nation-states south of the border.

Will America's big assignment succeed within northern Mexico where few indigenous live? Maybe. Will it fail in central and southern Mexico, and most of Latin America where many indigenous live? Yes.

Artificial world of economic growth

Throughout the region and especially in the common market for southern South America (Mercosur in Spanish and Mercosul in Portuguese) 'an artificial world of economic growth [has been built on the basis of] inefficient industries prospering behind a wall of trade barriers',[16] says Alexander J. Yeats, the principal economist in the international trade division of the World Bank. 'Trade among Mercosur members [Brazil, Argentina, Uruguay and Paraguay] surged to US $12.3 billion in 1994 from $4.2 billion in 1990. The fastest-growing items in intra-Mercosur trade are cars, buses, agricultural machinery and other capital-intensive goods that are produced relatively inefficiently in the four member countries. In other words, Mercosur countries, insulated from outside, are investing in factories that build products that are too expensive to sell to anyone but themselves.'[17]

This is heresy among U-World bureaucrats of the World Bank, the IMF and the Inter-American Bank. They are loath to decide whether such free trade pacts divert or create trade, because free trade is a political hot potato for all democracies in Latin America. Alexander's unofficial report on Mercosur could be the smoking gun that kills the idea of one free trade pact for all the Americas or the extension of NAFTA to other countries in Latin America, or both.

Mercosur

Mercosur is the end of the beginning for the countries of southern South America. They have agreed to give up their past, statist economic policies and try out enterprise capitalism through a common market, customs union and free trade agreement. This is their big assignment.

'As investors, both local and foreign exploit the potential for growth, scale and specialisation offered by this new market, this southern cone will be emerging as an agribusiness and mining superpower, with a diversified and modernized manufacturing industry fueled by cheap and abundant energy. This is the world's fourth-largest integrated market, after NAFTA, the European Union and Japan. The Brazilian state of São Paulo has now displaced the United States as the largest single outlet for Argentine exports. For an Argentine firm, Mercosur more than triples its domestic market; for a Brazilian one, the expansion is not even half [because] annual growth in Brazil is equal to all of Argentina's annual consumption.'[18] Americans should be familiar with this statistical phenomenon because annual growth in the US is equal to all of Canada's annual consumption – in cars, beer, clothes, computers, etc.

Although Brazil is Mercosur's dominant power, much of Brazil doesn't care about Mercosur. Infrastructure remains a crucial bottleneck because national rail lines are inadequate, roads are bedeviled by national customs procedures at each border and airports are without sophisticated cargo handling equipment. Mercosur needs less government red tape and more common rules – that is, if and when it will join NAFTA in a free trade agreement throughout the Americas.

Lesson from Mexico for Latin America

Here is what all of Latin American government agencies offer to the world's foreign investor: overstaffed and under-competent bureaucratic disaster zones and networks of red tape. 'Over-regulation of labour markets hits job-seekers. Employment taxes hit employers, workers and would-be workers alike. Speedy, honest and open civil justice is needed both to ensure the enforcement of contracts and so – perhaps even more important – the confidence that they can be enforced; in practice, nearly all courts are slow-moving and too many corrupt.'[19]

Mexico has shown that habits and institutions cannot be transformed overnight. A populist reversal of NAFTA and Mercosur is not imminent yet, but the malaise and unhappiness with the results from these free trade agreements is real. Also in many countries, policemen and former policemen engage in murder as the spiral of violence now consumes '13–15% of GDP on security expenses (both private and public). That is more than total welfare spending. It represents a crippling burden on the economy.'[20]

The most likely outcome to the plight of Mexico within NAFTA, the trade diversion efforts of Mercosur and the lack of serious socio-economic reform within the rest of Latin America is a long war towards the final victory of stable economic growth. Through the end of 1996 Mexico was standing in the middle of a stopped escalator on its way down. Even though Mexico was standing still it felt like the economy was still falling. On the other hand, the Mercosur countries were running up a moving down escalator, and they felt and knew they were standing still. The Andean countries have simply fallen off the escalator because they could not hold on as the escalator increased its speed going down.

Booms and busts

'All Latin American booms [during the nineteenth and twentieth centuries] have had in common the conditions of excess global liquidity resulting in capital in flows, and high or stable commodity prices. Whether the regime was pro-market or interventionist seems to have mattered less. [Again] Latin America is growing. Over the past five years it has had a large capital inflow – it took in $201 billion (in US dollars) of private capital in 1995 alone and $182 billion the year before. When large amounts of capital flow into Latin America, the result is an investment boom that causes a spurt of economic growth – growth, however, that has never been sustainable ... [because national] savings tend to accumulate in offshore accounts and behave like foreign hot money.'[21]

During the 1970s, 1980s and 1990s, there were no miracles in Mexico, Brazil, Argentine, Chile and elsewhere in Latin America. The political elites convinced international bankers that the former's new market orientation – that is, to reduce trade and investment barriers – could pull their countries out of poverty and underdevelopment. During spurts in economic growth, the savvy upper and upper–middle classes grabbed all the benefits and saved at a high rate in offshore accounts. 'The middle and lower–middle classes that suddenly emerged during boom periods lost everything in the bust'.[22] And the poor became truly destitute when the free market system collapsed from external shocks, such as a liquidity crisis or a fall in commodity prices.

Neither state intervention nor free trade schemes create the socio-economic conditions for long-term stable economic growth in Latin America. The fault lies in the political dynamics of each nation-state in which a few control all the resources and the rest have no stake in the economic system. America can do little to resolve the enormous problems facing contemporary Mexico and the other Latin American countries.

Nation-state involvement in the economy

Both cosmopolitans and loyalists believe in the power of the state. Their joint memories include an activist government role in the economy in which nation-states collectively seek to manage international trade (through the WTO/GATT) and the international flows of money (through the IMF and Bank for International Settlements). Their several memories include separate national decisions on interest rates and the foreign exchange value of the national currency. Both loyalists and cosmopolitans are on the verge of a new definition of economic sovereignty because everyone accepts government for what it is – the power behind the market, the final arbitrator of prices, values and outcomes.

What is the fraction of a country's income which the government spends? Let's look at several crucial points:[23]

'First and most important, the share varies widely. In America, public spending is 33% of GDP; in Germany it is 49%,' Singapore (20%), Sweden (68%). 'Behind these averages lie correspondingly large differences in underlying economic philosophy: in the size of the welfare state, in the extent of income redistribution through taxes, in industrial and labour-market policies, in ownership and/or regulation of natural monopolies, and so on.'

Second, 'since 1980 the public-spending ratio has increased, on average from 36% of GDP to 40%. National governments not only retain wide discretion over the extent to which they control resources, but after 15 years of accelerating integration are tending to control more, not less.'

Third, 'global integration has left governments with about as many economic powers as they ever had. [Yes] the international competition for capital is fiercer, [but] this is [not] exercising a new influence … on the design of all manner of government policies. The bulk of investment worldwide continues to be financed out of domestic savings.

'Lastly, once capital has been transformed into fixed assets, it is no longer mobile at all … and has to put up with government policy.'

These facts about the results of economic integration are agreed to by both cosmopolitans and loyalists. Where they differ is on the when and how governments might raise barriers once again against foreign goods. Rooted cosmopolitans would prefer national governments to do nothing to restrict trade and capital. Loyalists might opt for higher tariffs, quotas, strict labor codes, negative interest rates for foreign funds and so forth. This is a government's sovereign right to do whenever public well-being is endangered by free trade agreements, customs unions and common markets.

Here we are engaging in foresight. If we forecast a bad patch for the international and domestic economies, such as end to the very long-term 50-year business cycle, then we need to alter our future by making better choices. This forecast gives us no room to get emotional over Mexico and the Americas. Such a forecast may lead us to prefer the loyalist prescriptions for the national economy of the US.

If we forecast a good patch for the international and domestic economies, such as a continuation of stable economic growth, then we can make a leisurely choice between the U-World, or loyalist nation states, or combine the two views of the world. This forecast gives us some room to add Mexico to the big assignment of creating a free trade area from East Asia, across North America, to all of Europe. Such a forecast may lead us to prefer the joint U-World–loyalist prescriptions for the national economy of the US.

Let us assume that the good times continue. National governments are much more involved in the economies of loyalist nation-states. The former want the latter to be successful for the simple reason that national economic success keeps the U-World at bay. Again, the key to success is good political and socio-economic decisions by loyalist nation-states about their national economies and commitments to free trade within the international economy.

Costs and benefits

How should America weigh the costs of deepening its commitments to Mexico, adding new members (such as Chile) to NAFTA, and attempting to create a hemispheric free trade agreement against the benefits of stable economic growth in the distant future?

This too is a case of decision-making under uncertainty. Remember America has choices. One option is to offer Mexico more financial support for its efforts to raise productivity, increase efficiency, create high-paying jobs and establish new industries south of the border. If America chooses Mexico only, it does so because of the common socio-economic interests of Mexico with Texas, New Mexico, Arizona and California, and the need to assimilate Mexican immigrants into the American mainstream.

A second option is to offer Chile and one or more other Latin American countries membership in NAFTA. If America accepts Chile, it does so with the full backing of both Canada and Mexico, and the persistent skepticism of both the people and Congress of the US.

A third option is to create a new trade agreement for all of the Americas. If America seeks to resolve the problems of artificial economic growth in the Mercosur countries, and the negative growth in Central and South America as a whole it does so without the prospect of political success. A free trade agreement for all the Americas is a bridge too far.

Option one keeps the Mexican border areas calm, but incurs a cost in terms of financial subsidy to the ruling classes of Mexico. Could this money be better spent on free trade across East Asia, North America and Europe? Decision theory offers us a way to weigh the costs of doing nothing for Mexico, vs. doing something for Mexico, versus doing everything possible for Mexico.

Will Mexico beat the fate of Salmon Day? No. Will Mexico's failure drag America to the fate of Salmon Day? Yes if America pours its money into Mexico. No if America chooses East Asia and Europe for its trade and investment opportunities under free trade.

Decision theory offers us a way to see clearly our long-term problem with Mexico and how to organize ourselves to beat the fate of Salmon Day.

Salmon Day

Salmon Day is the experience of spending an entire day swimming upstream only to get screwed in the end. Bell Atlantic and Iusacell failed, and Telmex gained market share in the cell-phone segment of the Mexican telephone market. However, only Panamco won support on Wall Street for its Mexican and Latin American investments. Of course, the failed Mexican firms are victims of Salmon Day.

Here are some summary thoughts. First, the transmitters of America's common knowledge about the big idea got it right when they invited Canada to join a collective effort on free trade for North America, but they may have gotten it wrong when they invited Mexico to add its weight to this shared work on free trade.

Before the 1994 peso devaluation Mexico had an economy in tradable goods equal to the regional domestic product of Los Angeles, California USA. After devaluation Mexico's economy shrunk to the size of the local economy in the Watts section of Los Angeles, one of the most impoverished neighborhoods in the city. The economic reasons for free trade vanished overnight, but the political reasons remained paramount and of crucial importance to America's southern frontier with Latin America. Peace along *La Frontera* and in America's Latino communities was the price for the financial support of Mexico.

Second, the editors of America's common knowledge are in the process of coming to grips with the fall out of real, hard facts about the results of free trade under NAFTA. One of their vivid memories must be that NAFTA did not create more jobs in the US. Another is that Mexico's low-paying jobs did not destroy jobs in the US. Thus America should not take unilateral actions against existing free trade initiatives.

Third, managers and others in the professional class must decide whether America should deepen its commitment to Mexico. This is part of America's big assignment for the twenty-first century. If such a commitment takes nothing away from East Asia and Europe, then the risk of avoiding a swim upstream may be worth taking. However, to decide for Mexico, and against East Asia, or Europe, or both is fraught with the danger of swimming upstream and getting screwed in the end.

Canada and Mexico are neither political nationalizers or economic globalizers. They cannot use their nationalisms to strengthen their position around the world. This makes them a part of the larger US economy that is now known as NAFTA. In this way, Telmex and Panamco become the beneficiaries of the American market-rationale approach towards the international economy.

Wisdom: decision theory lays out possible choices. Pay attention: the choice for Mexico only leads to Salmon Day, but the choice for Mexico, Europe and East Asia can lead to a second American century. Open the doors: America reduces its commitments to Mexico and deepens its commitments to East Asia and Europe; America makes a choice to defend its most important interests.

What do I get?

Here is our to do list:

- Buy into America's limited role in Mexico.
- Accept the commitments to East Asia and Europe.
- Build on the memories of partnerships with Europe and Japan.
- Reimage ideas to include Japanese and Chinese themes.
- Recognize Mexican immigrants as 21st century citizens.
- Create business ties among Asia, North America and Europe.
- Use decision theory to weigh costs of the Mexican option.

Remember these words: Salmon Day is not inevitable for America. The US must stay away from an emotional commitment to Mexico that is all out of proportion to what the Mexican economy can do to advance stable economic growth in the US.

Chapter 3

The West

'We run our fingertips along trend-bumps as they speed past. We Braille the culture.'[1]

Faith Popcorn, professional trend spotter.

'A lot of ordinary Asians, like their governments, are sniffy about "Western values". Should a good Asian embrace Mickey Mouse? It is a fair question, which is why the provisional answer is so interesting. So far, most Asian theme parks owe more to DisneyWorld than to Confucius.'[2]

The Economist.

America in the making of the world

Americans, in the first two centuries of their history, saw Europe in the east as a different civilization, Orthodox, Communist or Moslem in religion, a profoundly tradition-bound conservative society. This is not surprising if one recalls that what Americans call Europe or 'The West' itself became defined only later as the Normans, Portuguese, Spanish, English, Dutch, French and some Italians conquered the Greek lands in Sicily, the Americas, the Japans, Cathay (China), India, the Arabian states, and sub-Saharan Africa. Where do Europe and 'The West' end? They stop where local societies in eastern Poland, Ruthenia (the tail of pre-World War II Czechoslovakia), Galicia (once Austria, then Poland and now Ukraine), Wallachia and Moldavia (pre-World War I Romania), and the Turkish settlements in southern Bulgaria and northern Greece do not benefit from the Renaissance, Reformation, industrialization, democracy and the international economy.

In America's rush to drive the French, Mexicans, Russians, British, Native Indians and Spanish from North America, Americans lost sight of 'The West.' They transmuted Europe into a stereotyped view of a fanciful English royal family and aristocracy. This is not to say Americans were completely uninformed or always misinformed about the Europe of reality. Nevertheless, in the American popular imagination, the mythical Hun and the real German often remained undifferentiated, even some of the most general cultural, linguistic and historical memories were not settled upon until after World War II.

What were the drives, which first impelled American business executives to push out into the unknown worlds beyond the mass market of the US? How did the expansion of global enterprise maintain its momentum until no countries were without Coke, McDonald's, Hollywood and CNN? We need to study America's icons of memory, its past achievements and its traditions. Then we need to destroy them as America sets out on the path of reimaging itself for the world.

The expansion and resulting rise to world predominance of the US meant the obscuring of the brilliant cultures of Europe and Asia by the newer and more dynamic global, largely American life style. Not until recently were some European countries, Japan and a few South East Asian countries able to mount a consumer revolution of their own comparable to that which transformed America in the twentieth century and gave to America its decisive lead in business, management and marketing.

Our big assignment in this chapter and throughout the rest of the book is trend spotting. What is faith? What is popcorn? Whither America's role in 'The West'?

Icons of memory

What are America's icons of memory? They represent faith in America's ongoing work both at home and abroad. In the previous two chapters, we discussed the following icons.

Culture

America promotes trends suffused with tradition. Americans ache for the lost past, but they work hard on the promising future. If Americans have to make a choice between old and new, they opt for the deconstructed idea, reformulated product and upgraded life style. America sees its worldly task as the pursuit of the global, largely American life style irrespective of whether others are ready for free trade in standard American English, the Internet, management theory and mass marketing.

Language

America seeks converts to its way of doing business and steadfastly refuses to accept the idea that other languages harbor valuable conceptual resources. The frame of reference is English. Americans work within the folk tradition that everyone who counts (that is, everyone who wants to do business with the US) knows and speaks English. America sees its worldly task as the pursuit of non-native English-speaking people into its orbit of fame and riches.

Society

America wants the hearts and minds of everyone who lives within the heartland of North America. Americans want Canadians and Mexicans to resemble them and nestle within the youthful culture of the US. If Americans must, they differentiate between English- and French-speaking Canadians, and they accept the revelations of sensational differences with Mexicans who live in Mexico or that country's 'near abroad', California and Texas. America sees its worldly task the pursuit of its own socio-economic truths at the expense of others who might share the same geographical space.

Immigrants

America wants everyone who comes to the US to convert as fast, and much as is humanly possible, to English and the American way of life. Americans want to be the world's village in which everyone shows a preference for American culture. America seeks its worldly task as the pursuit of souls yearning to feed on the democracy of the federal–state partnership, the internal checks and balance, and the Bill of Rights.

History

America wants everyone to forget the past. The loss of unique ideas; the reduction of language; the collapse of class and generational differences; the seduction of immigrants: America wants to conquer memory. America's worldly task is to put its commercial icons on the shelves of the world so converts can preach the new universal America faith and venerate its living saints, free international trade, foreign direct investments and global business enterprises.

Americans believe in American exceptionalism, and that the US is unique among nations. People came to the US to forget the past, their culture, languages, history and traditions. People came to the US to gain the future – namely, Hollywood-style entertainment; TV English; fast foods and fast cars; and the snap, crackle and pop of instant satisfaction. People came to the US to go through a systematic transformation (a *Zeitgeist*) of their socio-economic values. These foreigners became Americans, the most loyal of loyalists to the American republic. They too are the keepers of America's icons of memory.

Football traditions within North America

Within North America, the US is the most important country by its compelling history, overwhelming economic strength and its preference for individualism. Some of its traditionless trends – namely, the ritual of eating snack foods and drinking beer while watching football on television – are shared with Canadians. A few American traditions become one more American trend in Canada that seems to be replacing Canadian traditions. Here is one story.

Professional North American football (not FIFA soccer football) in Canada goes by the moniker CFL. It has a longer football field, fewer downs and its own 'world' championship game, the Grey Cup. In 1995 an American CFL team, the Baltimore Stallions, won the Grey Cup in Alberta, Canada; in 1996 all US-based teams were gone from the CFL because they did not draw the fans needed to fill the stands in the US; and in 1997 virtually all CFL teams in Canada were on the verge of folding because they too could not fill their stands in Canada. Average attendance slipped to 22,000 and only Winnipeg had a profit. Not even Doug Flutie, the American college player who gave football fans the Hail Mary Pass and played for several CFL teams, could save the CFL from the NFL, professional football in the US.

The NFL plays big football. Its players are around 300 pounds rather than the average 200 pounds in the CFL. The NFL fills its stands with well-fed, pampered fans, some who drive from Vancouver and Toronto in Canada to San Francisco and Buffalo in the US to cheer their favorite teams on to victory. Others fly to the NFL's world championship game, the Super Bowl, to soak up the sun, drink Molson or Budweiser, and be a part of North America's favorite pastime, Sunday afternoon and Monday evening NFL professional football. The American trend of the 1960s became a tradition in the US in the 1970s, a trend in Canada in the 1980s, and a tradition in Canada in the 1990s.

Gray markets

Ask most Americans about this bit of Canadiana, CFL football, and they will give you a blank stare or plead ignorance. Ask Canadians and they will blame the better game from the NFL, the inadequate performance of the owners and management of the CFL and the gray market in satellite TV service from the US. Their answers are both worried and informed while Americans do not know or could not care less. Just give them the popcorn of NFL football and American trends will become traditions both in the US and Canada. Canadians shrug their shoulders because they know they are unable to stop the forces of socio-economic history from across the border.

Canada's history is different because the country began without a revolution to throw out Britain, continued as a compact between two founding peoples (English and French) and took in different foreign people (for example, South Asians and Hong Kong Chinese) to populate its land. Americans view English-speaking Canadians as first cousins who prefer to live in a different house. Canadians see Americans as second cousins who eat too much, brag too much, play too hard, and upset the peace and tranquility preferred by English, French and new Canadians.

If Canadians can get away with a bit of nationalism, they will try Crown Corporations or Canadian-owned private corporations to resolve economic problems. State enterprises failed in the petroleum and gas industry because the 'Seven Sisters' were too powerful in both the American and international markets for the Canadian nation-state. Environmental mandates for privately owned Canadian firms also failed in the pulp and paper industry as large Canadian-owned firms put 40–50% of their assets into the tree farms and mills of the southern US. Many became for all intents and purposes firms headquartered in the US.

However, government-directed initiatives succeeded in the telecommunications industry because the telephone and satellite firms bought into the national argument that Canada's culture is in danger of being overwhelmed by cable feeds from Hollywood, CNN, HBO, PBS and many other sources in the US. In 1997, American satellite TV programming reaches over 90% of Canada's residents. 'Otherwise law-abiding [Canadian] consumers openly bought and installed satellite dishes the size of garbage can lids and subscribed to pirated services that bring in Seinfeld, The Simpsons and other TV shows from the United States.'[3]

The fault lies in the inability of the governments of the US and Canada to sign an agreement that lets TV services from the US to come openly into Canada and vice versa. 'Washington wants the same type of unfettered access for satellite signals as for auto parts and other free trade products. For cultural reasons, Ottawa insists that at least half the programming on a satellite service be Canadian, even though Canada so far has been unable to offer its own satellite service. So for now, there are no legal satellite services in Canada. But some 200,000 Canadians pay for programming from the United States that the [Canadian] government has warned is illegal.'[4] Some of this programming is NFL football and American sitcoms. However, a good part of it is for those straight-laced Canadians who are willing to stop acting like Mounties and wink at the law to get PBS programming in Hamilton, London and other cities in southern Ontario.

TV memories

Around the table at Mexico City's Hacienda de San Angel, some Mexican business executives talk about professional football from the NFL in Mexico and Monterrey. Exhibition games have been played in Mexico from time to time as they have been played in England, Ireland, Germany and elsewhere in the world. However, those Mexicans with satellite TV service usually see their NFL football games from TV stations in Houston, Denver or Los Angeles.

Mexico is indeed a different country with a language and culture that sets it apart from the US. Americans think of Mexicans as difficult neighbors who must live in their own homes, work in their own factories and live their lives according to a collectivist meso-American culture. Americans do not think of Mexicans as part of America, Europe and the West.

Mexicans see Americans as difficult neighbors, too. The latter are too close to be shut out, too successful to be forgotten and too demanding to be ignored. Mexicans fear the pull of Hollywood and CNN, but they cannot help themselves as they are drawn to play at Disneyworld and shop in New York. Mexicans think of Americans as their evil other side, the victors in the never ending battle between tradition and trend, faith and popcorn, and good and evil.

After all, what good can come from a country that prefers NFL football to FIFA soccer, say loyalist Mexicans. At the House of Tiles, Sanborn's restaurant near the Zócolo, you hear the table talk about the US:

> *'The* gringos *hosted World Cup. Yet they let the Brazilians win. And they still don't have a professional football league worthy of inclusion within FIFA.*

'The gringos *bought the World Cup just like they bought northern Sonora and attached it to Arizona, and they stole the rest of northern Mexico, the gold in California and the oil in Texas.*

'Now these gringos *want are souls, too!'*

Recently, Washington and Mexico City signed a satellite TV agreement 'to keep content restrictions to a modicum, a word expressing moderation in both Spanish and English. The United States expects Mexican broadcasters to reserve 7 percent of air time for public interest broadcasting, perhaps in English. Mexican broadcasters [will enjoy] full access to the lucrative Spanish-speaking market in the United States.'[5]

Mexico is protected by its Spanish-language barrier. Canada has no such barrier for three-quarters of its people who speak English as their native language. Hence, Canada forgoes market share and works very hard at protecting its cultural institutions from the popcorn that might come in from the US. Mexico accepts the popcorn. Mexico works very hard at suffusing some American trends with its own meso-American traditions.

Professional trend spotting

Faith Popcorn, a professional trend spotter, says people 'Braille the culture', meaning 'We run our fingertips along trend-bumps as they speed past'.[6] In Mexico, sales of Bart Simpson *piñatas* soar as sales of traditional donkey *piñatas* fall. Current events from the US are news in Canada as the ups and downs of the Whitewater problems facing the US President and First Lady loom larger than the Canadian Prime Minister's pre-election talk about repealing the general services tax (or GST).

Things are moving so fast in North America and Mexico that the paraphernalia from this year's Super Bowl winner is reduced after the first quarter of play, given additional discounts at half-time and remaindered when the game is over. NFL football is a U-World culture built almost entirely on a preference for passive participation in sports entertainment. It's a trend that became a tradition in two loyalists nation-states, and through NAFTA is seeking to make in-roads in a third loyalist nation-state as well.

All trends seem to be equal in the eyes of Americans. Their trends must have brevity and sensation, a bit of recklessness, and youth. Americans who are in their sixties want products that cater to their youthful fifties; Canadians who are in their fifties want goods that cater to their youthful forties; Americans who are in their forties want services that to cater to their youthful thirties; and so on. The Beats begat the Baby Boomers who begat the Punks who begat Generation X, and all want goods and services that are one decade younger than their chronological age. Now some Mexicans, Europeans, East Asians and others are becoming trend addicts, as frenetic over changing ideas, sales and marketing techniques, and postmodern management practices as are Americans. 'The trendiest trend in culture right now is trend-spotting.'[7]

Trend-spotting Christmas

According to Earl Wendell Count, a professor of anthropology at Hamilton College in New York, the Christmas story goes back to 'wealthy Babylonians exchanging gifts in an empire that flourished 4000 years ago'. There were people of pre-Christian Europe, who believed they could banish evil with decorations of winter's fir greenery and its promise of renewed life. Mistletoe had its place in Druids' worship and old Aryan legends. The Yule log and bonfires to dance around were traditional to many civilizations.

> *'The church fathers, the bishops of the new religion, took the merriment, the greenery, the lights and the gifts from Saturn and gave them to the Babe of Bethlehem. It was the common folk who, out of deeply felt faith, inspired the celebration known in modern times. Christmas became great because men and women have given it their deepest and most enduring selves.'*[8]

The story of the Christmas holiday explains a great deal about how death does not come easily to a good idea. Forget the birth of the Jewish child in the Roman province of Palestine some 2000 years ago. It didn't happen in December. About 300 years after the event, the newly established Catholic

Orthodox Christian Church decreed the birth would be celebrated in conjunction with the pagan ritual welcoming the Sun's slow return to warming the Earth.

In those days, a bishop (of Myra, to use the modern Turkish name) gave gifts to help young girls stay out of slavery and away from child prostitution. He became known as St Nicholas in Asia Minor and Greece. Much later he was called Father Christmas in England and Sinterklaas in The Netherlands, and then jolly ole St Nick in the former English and Dutch colonies of a newly independent US.

The nineteenth century was not an easy time for him in the US. During the holiday, many people got drunk and debauched themselves. Traditional Christmas among the highly religious Calvinist Protestants was one of public rectitude, alms for the poor and work when the 25th fell during the six-day work week. Early in the twentieth century alms for the poor were forgotten as gifts for one's children became the norm. Catholic and high church Anglican ceremonies made Christmas a religious pageant with fasting on the eve, Midnight Masses and presents from St Nick opened by children around the family's Christmas tree.

Trend-spotting Christmas in contemporary culture

Then in the 1930s Coca-Cola ran ads showing St Nicholas–Papa Noel–ole St Nick as Santa Claus drinking Coke after his hard night's work on Christmas Eve. From then on the world was filled with the sleigh, toys, reindeer, Rudolf, holly, bells and everything one could think of to make Christmas a commercial extravaganza. As the Dutch become less religious, traditional Sinterklaas with his book of good and bad deeds is in decline while Santa Claus with his gifts from around the world is taking over all Calvinist, Catholic and Moslem holiday celebrations. In the Zócolo, the seat of Aztec, Spanish and Mexican power in Mexico City, the larger-than-life Mexican flag is surrounded by thousands of lights depicting all of America's Christmas symbols. All children (American, Mexican or whatever) whether they be Christian, Jew, Moslem, Buddhist or atheist must have a tree under which Santa (or some other holy person) can lay his gifts. Today, in the far corners of the world, Christmas shopping and gift giving (in Tokyo's

Ginza, Manhattan's upper East Side and Istanbul's Phanar) are contemporary not religious for women who are in kimono or Western dress, who are veiled or unveiled and who wear different things at home, in shops and at the office.

America's business loyalists are a proud lot. Their view of the Christmas story now engulfs the world. America imports new Christmas tree decorations and toys for children; and it exports more expensive tool kits, up scale fashions and seasonal foods to customers. Although the world market is not a cultural concept, the contemporary commercial market for Christmas cheer is a cultural concept. Similar market segments exist all over the world; of course, the goods and services that fill up these segments and niches differ from place to place, but they too are becoming more similar as the world accepts global materialism over localism. As a consequence, America's proponents of the U-World see the contemporary Christmas commercial saga as the way to convert loyalists from America and the rest of the world.

Recycled nostalgia

Both rooted cosmopolitans and loyalists have memories. Each group works on the other. Perhaps, one will convert the other, provide themselves with joint and several rememories, and turn all of their commercial work into shared remembrances. Let's see how this might work out as trends become recycled nostalgia and endure as traditions within the Slavic lands east of Europe.

For example, Russia is racing to catch up with changes in American tastes, European styles and Western ideas, learning new names and constantly seeking to find newer ones. Russian consumers want to be part of the action, ride the crest of the waves instead of floating helplessly while the action is elsewhere. Western culture is thrown about on screens, billboards, in concert halls and art galleries, filtered through thousands of minds trained in America and Europe, executed with millions of sales and marketing gestures. Much of the socio-economic interactions between Russia and the West is a mess because some things are hot today and gone tomorrow.

Everything within the center of Moscow is a sign of what Russians want to be and where they are going. The hunt is on for reliable trends without the presence of the past. Russians believe traditions from the last 70 years will kill politics, economics and culture. This prophecy is self-fulfilling, and it is unrealistic for Russia. The communists succeeded in eliminating the monarchy, the Russian Orthodox Church and the landed aristocracy from public life.[9] Russians believe traditions from the previous 400 years will restore their blessed past. This prophecy is also self-fulfilling and unrealistic for Russia.

Tomorrow's Russia will be a little of its combined Mongol–Czarist–Soviet past, some of the in-between times of the present, and a lot more similar to the cultural stranger called the Western world. Americans are really proud of the shrinking world they created, but Russians fear losing control to foreigners – especially, Germans, Americans and Japanese. The battle for the Russian soul has just begun. This is the realistic prophecy for Russia.

Trend-spotting contemporary Christmas in Russia

During the Christmas season, Russia celebrates four separate days of exchanging gifts:

- December 25 is Catholic Christmas, the day for Santa Claus, from the new calendar of the Western Latin Roman Catholic and the Eastern Greek Orthodox churches.
- December 31 is a holiday of the former Communist government, the day for Father Frost and Snow Maiden, and New Year's Eve according to the new calendar.
- January 7 is Russian Orthodox Christmas, the day for the Christ Child, from the old, pre-Soviet Union Julian calendar.
- January 13 is New Year's Eve according to the old Julian calendar.

> *'On billboards all over [Moscow and St Petersburg], Santa Claus toasts the holidays with a giant bottle of Coca-Cola. At the Sadko arcade [in Moscow], a mall that caters to Russia's new rich, wealthy shoppers*

> *have been snapping up $2,000 artificial Christmas trees (in US dollars) and $700 dacha doll houses. Small shops and even makeshift vodka kiosks are festooned with blinking Christmas lights, polyester Santas and plastic angels.'*[10]

Here is what is happening in Russia's largest cities. Older Russians recycle some of their ancient holiday traditions. Younger Russians borrow many of their newer holiday customs from the West. The newly rich Russians decide not to work for at least a month during this very long Christmas–New Year holiday period.

Iconoclasts

In this way, Russians are destroying older icons and becoming iconoclasts. Younger and newly rich Russians are going through a process of reimaging themselves – some aping the West, Europe and America; and others preferring the East and pan-Slavism. Most of these Russians are loyalists to the new Russia. This is the country that emerged from the rubbish of the former Soviet Union, the long-lasting Euro–Asia country in which the people now have a choice:

- about one-third of the people retain their nostalgia for the traditions of the former Communist government;
- another one-third favor the trend towards a new democratic government; and
- one-third prefer pan-Slavism as a counterweight to the West, its Renaissance, Reformation, industrialization, democracy and the international economy.

The destruction of Western icons is a very old tradition in Russia. If Western ideas, such as Marxism, Communism and Capitalism, corrupt Russian values, customs and traditions, Russian iconoclasts believe it is better to destroy these icons, too.

Today, Russia is searching its memory for its true past as a guide to what it wants to be in the future. Can Russia change? Many historical factors work against democratic capitalism in Russia: xenophobia, absolutism, indifference to property rights, contempt for the weakness of Western liberalism, and a 'yearning for all-embracing ideologies rooted in the anti-intellectual and eschatological style of the Russian Orthodox Church'.[11] Something unfamiliar, something altogether undreamed-of in the West might come along to lead Russia out of its present morass – a rough mafia bear, a horde of speculators with icons and axes, or a resurrected corps of entrepreneurs and business executives.

Win–lose win–lose world

The U-World is at the beginning of a mixed bag era for trade and investment opportunities in Russia and its 'near abroad' neighbors in Europe and Asia. Some will be important successes and help Russia avoid Salmon Day, and others will be spectacular failures and drive Russia into Salmon Day. Remember: 'Russia's economy is roughly the size of South Korea's, and barely a 20th the size of America's'.[12] Let's look at a few examples and see who is involved from the West and East Asia.

Caspian Sea oil

Mobil, Chevron, BP, Total, Elf, Petronas, China National Petroleum Corporation and other oil companies are in the Great Game of trying to exploit the oil and gas reserves from Iran, Turkmenistan, Kazakhstan, Azerbaijan and south-west Russia – all countries with territorial waters in the Caspian Sea. This region may have as much as 100 billion barrels of oil compared with Kuwait's proven reserves of 97 billion.[13] These firms are trying desperately to replace declining fields in Alaska, the North Sea and elsewhere in the world. The oil majors have vast capital resources to invest, but they do not have the pipelines built through south-west Russia, from Azerbaijan to Georgia, or, perhaps, through Iran. Their future Caspian gushers are bedeviled by high political risks.

Here is a short list of the risks involved. How much of the deal should be reserved for the state sector in Russia and the former Soviet Union?

How much of the oil can be put on tankers, trains and trucks in lieu of a pipeline? Which pipeline route will be acceptable to most of the Caspian Sea nation-states? What law will apply when one or more of the nation-states changes the exploration deal unilaterally or takes over foreign assets? If these risks become too high, then Exxon, Mitsubishi and the China National Petroleum Corp. might build an alternative 13,000-mile gas pipeline for about $22 billion (in US dollars) across Kazakhstan, Turkmenistan, Uzbekistan and China to Korea and Japan.[14]

Energy-hungry East Asia prefers the Central Asia–China–Japan pipeline. Japan, for example, believes its gas needs will double by 2010. Korea and China also forecast doubling and tripling their gas needs early into the twenty-first century. At the moment, no oil and gas exploration projects in East Asia show the promise of the reserves already found in Central Asia.

Politically, all of Central Asia wants to get out from under the thumb of Russia. However, none of the former Soviet Union countries know how Russia will react to large-scale American–Chinese–Japanese influence in Central Asia. This is the peril of the Great Game in Central Asia.

Automobiles

Ford, GM, Volkswagen, Daewoo and Kia are setting up kit assembly plants in Russia and the 'near abroad' countries in eastern Europe and Central Asia. Their automobiles cost a third to a half more than the Russian-made cars, and the former are the target of car thieves and the police.[15] Although Russian demand is high for the roomier American-style cars, Russia lacks networks of parts suppliers, retail distributors and gasoline service stations. So those with money still buy Ladas as their second and third cars for use in the big cities. Moreover, average disposable income in Russia is much lower than in east Europe, China and East Asia so those without a great deal of money also buy Russian-made cars. Hence, foreign-owned automobile investors are putting a great deal more money into the emerging markets of Poland, the Czech Republic, Hungary, China, Thailand, Indonesia and India.

Small businesses

Small businesses produced 12% of all goods in Russia in 1995. Many are owned by women. They are funded by a $300 million revolving fund of the European Bank for Reconstruction and Development; First Woman East, a London-based venture capital fund; and the US Russia Investment Fund.[16] These firms produce new fashions for women executives, language schools and training centers, media and magazines for the energy industry; FM radio stations, donuts, photo labs and many other goods and services. These companies are small beginnings in the quest to bring market economics to Russia.

Here are some of the valuation problems: 'On an asset basis Russian companies always look incredibly cheap. On a production basis they still look quite cheap. On a price to sales basis they begin to look like they might be priced about right. But on a p/e basis, taking account of corrected earnings, they all look blatheringly expensive,'[17] says James Nail, head of research at Deutsche Morgan Grenfell's Moscow office. Foreign investors tend to be blind to the real business facts. However, the direction of the Russian market in general, and sales and profits for many Russian companies seems to be up in the bigger cities of Russia.

'What needs to be done?' Settle the path of the oil and gas pipelines. 'What do we get?' More automobile customers. 'What can we do?' Invest in small and mid-size businesses. 'Tell me what you are going to do on Monday that's different?' Bet on a long-term horizon for Russian investments, and look at all Russian sales and profit data with a very skeptical eye. Today, memory is best served by combining Russian desire for better goods and services with existing capital resources from the West and East Asia.

Go East

America may be over-reaching itself by trying to convert the former Soviet Union and the Moslem Levant to mass marketing. The people in the East (from the pan-Slavic Orthodox communities in St Petersburg, Minsk, Kiev, and Sofia, and the Sunni and Shiite Moslem communities in Istanbul, Cairo, Damascus and Tehran) are profoundly conservative. Their icons of memory differ greatly from those of North America and western Europe. Here are some examples:

Culture

Both communities favor tradition over trends. The premises, style and subject matter of past imperial and religious ambitions shape their frames of reference. Russia, Turkey and Iran see their regional task as the pursuit of competitive position in the 'near abroad' of Central Asia and the Arab Middle East. All have difficulty accepting the global, largely American life style.

Language

Both Orthodox and Moslem communities prefer their own imperial languages to English. Russian is losing the battle for eastern Europe to the resurgent commercial *lingua franca*, German, and for Central Asia to re-emerging native languages, Turkish and Farsi. Also older Russians prefer French as their second language. Since Arabic is the language of Islam, it sweeps across North Africa, the Middle East and much of the middle part of the old world. However, English is an important second language in Egypt, Saudi Arabia, the Gulf States, Pakistan, India, Malaysia and Indonesia. Many have difficulty with the idea of pursuing their fame and fortune in an English-speaking world.

Society

Both communities work within their own folk traditions. Neither share America's quest for immigrants as the mortar for a new people. Although Russia built new communities in Siberia and the Far East, the Turks in Anatolia and the Balkans, and the Arabs in Morocco and Iraq, both Orthodox and Moslem communities fought with the West over whose socio-economic truths would prevail in Europe, Africa and Asia. The fall of the Berlin Wall in 1979 was as great a defeat for Russia as the failure by Turkey to conquer Vienna in 1653. Their history is a contest with Britain, Europe and America in which over five centuries the West wins the cultural wars.

These two societies feel they are being culturally assaulted by the more successful global, largely American life style. Some of the symbols of American marketing power might even trigger a backlash against American shopping centers, exports, software, football, business and all the other aspects of the Americanization of the globe. The attempt to include the Orthodox and Moslem communities within the West's orbit of commercial success runs a high risk of Salmon Day.

Look to East Asia

Some thirty years ago pessimists would have reached the same conclusion about the ability of America to include Japan and the Chinese communities throughout East Asia and the Pacific into the West's orbit of commercial success. These forecasters would have said: Salmon Day.

Yet today Apple Macintoshes, McDonald's hamburgers, Levi's jeans, Cross pens, Friskies cat food, Coke's cola drink and Mattel's Barbie have made deep penetration in the elusive Japanese market. The same is true among the Chinese communities in southern and coastal China, Hong Kong, Taiwan, Thailand, Malaysia, Singapore, Indonesia and the Philippines.

The biggest change in Japanese, Korean, Chinese and Thai tastes is that their consumer cultures are more contemporary, more cosmopolitan, a creature of the borderless world favored by American advertising, sales and marketing, and business communications. Today, the main customers for up-scale products, such as theme parks, are the Asians themselves rather than expatriate Europeans and Americans who live in East Asia.

Asian values and the rootless generation

The leisure boom started in the early 1980s among the Japanese. Now other

East Asians (including many Chinese for whom the five-day week became common only in 1994) have money and spare time to enjoy themselves away from home. Tourism is expanding at a breakneck pace with 4.5 million Chinese going abroad in 1995, and with tens of millions of Asians traveling within their own countries.[18]

American or Western experience is not necessarily a good guide. 'A lot of ordinary Asians, like their governments, are sniffy about "Western values". Should a good Asian embrace Mickey Mouse? It is a fair question, which is why the provisional answer is so interesting. So far, most Asian theme parks owe more to DisneyWorld than to Confucius.'[19]

These Asian theme parks started out as shopping centers, like alien spaceships in congested city centers and undistinguished suburbs, air conditioned, places of fun for large crowds of people. Here are three examples in Bangkok, Thailand. The Ma Boon Krung or MBK duplicates the shops in Siam Square across the street, closes them in and ties them together with fast-food courts, ethnic restaurants, cafes, bars, supermarkets, department stores, hotels, movie theaters and amusement parks. Also the Seacon Square Mall has a roller-coaster, a Ferris wheel and a monorail. Moreover, the Leoland Central City Mall has many different types of swimming pools and a 'Traveltron' simulator theater.

Bangkok, Kuala Lumpur, Singapore, Jakarta, Hong Kong, Shenzhen and, of course, Tokyo are all competing for the dollars from tourism. One of the cities in South East Asia wants to lure Disney so that East Asia will have a second Disneyworld a bit distant from the one outside of Tokyo. If they cannot get Disney, they might settle for a Warner or a Universal amusement park. Or they might build one of their own with interactive computer activities that shed some light on the region's culture, language, society and history.

Governments throughout East Asia are worried about the results of rapid economic growth: a rootless generation without a good grasp of Confucius or Asian values. They do not like what has happened to many Japanese youth who seem detached from the past. They do not like what has happened to Chinese youth who come to Shenzhen to share in 'China's version of Miami, complete with restaurants, night clubs, resort hotels, glitzy apartments and a growing supply of gangsters. Millions flock here to enjoy themselves each year. [Most to play the horses and gamble, others to drink in German beer halls, and many simply to sin with leggy East European blondes.] Four fifths are mainland Chinese. Most of the rest come from Hong Kong and Taiwan.'[20]

All these themes are crowd pullers among secular Japanese, Chinese and other East Asians, and are shared by secular Russians, Germans, Americans and other Westerners, too. These themes are anathema among the devout and religious of East Asia, the West and, of course, the Moslem world. Unfortunately, adult entertainment, children at play, shopping as fun and having a good time are themes that condemn America, Europe and the West in the eyes of those it seeks to convert to modern mass marketing within a global economy.

Mirage of Immortality

If America is in political decline, it follows its cultural appeal is in decline too. If both America and Europe are both weakening, it follows the West's cultural attraction is weakening too. Today, the Chinese, Islamic and Orthodox worlds are all more self-confident. After four centuries of Western dominance, they are on the verge between something and something else.

Perhaps, they will come into conflict with the West. The fault lines already exist within the former Yugoslavia between the West and the Orthodox, in the Balkans between the West and Islam, and in the Pacific between the West and the Chinese.

Perhaps, one or more of these civilizations will adapt themselves further to some of the unique qualities of the global, largely American life style. Some of this is happening in both the Chinese and Orthodox worlds.

Perhaps, as in the distant past of the Arabs and the Crusades, the West will learn from these three traditional civilizations. This outcome is harder to predict because it requires several assumptions about cultural conflict, the applicability of rights-based individualism to non-Western societies, and the relative decline of the West.

The global, largely American life style only touches a tiny elite around the world. The rest of the people ignore American culture, society and history. Some even hate the West. Yet the intrusion of the market economy, management discipline and marketing techniques create wealth and increase consumption – something that is abundantly clear throughout East Asia. The following are the phrases used in Asia to summarize the drive towards modernization without westernization:

- *Ti-yong* (Chinese learning for the fundamental principles, Western learning for practical use).
- *Woken, yosei* (Japanese spirit, Western technique).

Within the Islamic countries of the Arab world, the import of intangible social and cultural institutions can be deadly – just ask the Shah of Iran. Modernization within the Moslem world must take into consideration that Islam is not just a religion but a way of life. Similar dangers face modernizers in Russia and the Slavic lands to the east of Europe, the new countries of Central Asia and the recent economic 'Tigers' in South East Asia, such as Singapore, Malaysia and Indonesia, and the Mexican borderlands with the US.

Notwithstanding the success of America, Europe and the West nothing about marketing prowess, language dominance and cultural strength is immortal. The same can be said about Japan, China and other great civilizations of the world.

America's big assignment

Both cosmopolitans and loyalists reimaged America to suit its new position in the world. They deepened America's commercial success in Canada and Mexico, the European Union, and in Japan and East Asia. They are extending America's marketing influence to southern South America, eastern Europe, the Middle East, South Asia and sub-Saharan Africa.

The saga of Mexico and Latin America is noteworthy because it shows how little Americans know about countries near to and cultures somewhat similar to the US. Before Americans expect a lot from the Orthodox and Islamic worlds the former will have to know a great deal more about the latter – that is, their culture, language, society and history, or their icons of memory. When big money is at stake Americans spend the time to learn about different people and cultures; we can pinpoint the beginning of our success in Japan, China and South East Asia from the willingness to learn how to do business and prosper in these foreign lands.

Perhaps America's really big assignment is to join with Europe in a combined effort to expand free trade and direct investment among the North Atlantic countries. NAFTA and the EU might make good partners in extending the commercial success of the West to those outside its reach in southern South America and in eastern Europe. More on this assignment in Part 2.

Perhaps, America's other really big assignment is to join with East Asia in a combined effort to expand free trade and direct investment amount the Asian–Pacific countries. NAFTA and APEC might make goods partners in taking the best from the West and the East and adapting it to those outside their joint reach in South Asia. More on this assignment in Part 3.

America is on the verge of something to something else. Let's see how these opportunities might work out.

Costs and benefits

How should America weigh the costs of broadening its links to eastern Europe in general, and specifically its ties to the Baltic countries, Russia, the Balkans and Turkey? Should America encourage a loose linkage for some and a tight integration for others in an expanded EU. Are some of these countries candidates for a new North Atlantic Free Trade Area? What will be the impact on Mexico, the Mercosur countries in southern South America and the other Latin American countries?

How should America weigh the costs of expanding its influence in the Moslem countries of North Africa, the Middle East, and of South and Central Asia? Are deeper economic links with both the Orthodox and Moslem communities simply a bridge too far for America given the immediate need to deal constructively with Mexico, Europe and East Asia?

These are all cases of decision-making under uncertainty. One option is to let western Europe to take the lead in integrating the Slavic and Turkish lands to the East, and America to take the lead in integrating the Arab lands of the Middle East. If America encourages Europe to be the more important partner in their collective relationship with the East, it does so because Europe (especially Germany and France) is in a better position to convert eastern Europe to a fully functioning market economy. If Europe encourages America to be the more important partner in their collective relationship with the Middle East, it does so because the US is in a better position to convert the Arab nations to a fully functioning market economy. This option is a bridge too far.

A second option is to offer selected Slavic and Moslem countries full membership in the EU and a proposed North Atlantic Free Trade Area. If America encourages Europe to make choices, those left out will want to know what the timetable is for their inclusion, too. This option is a bridge too far.

A third option is let the World Trade Organization (WTO/GATT) do the hard work of bringing one or both of these areas into the international economy. If America and Europe make this choice, they will do something that is within the art of the possible. Option three offers a way to avoid Salmon Day in two vital areas of the world.

Salmon Day

Salmon Day is the experience of spending an entire day swimming upstream only to get screwed in the end. All foreign investors in Russia seek market share without a full and complete set of information about costs, sales and profits. Western-style accounting reports are not available. Many foreign investment deals in Russia might fall victim to Salmon Day.

Here are some summary thoughts. First, the transmitters of the West's common knowledge about the big idea of a joint interest by Europe and America in the Orthodox and Moslem communities in the East got it right when the former decided to go slow in repositioning the latter within the international economy.

Second, the editors of the West's common knowledge (that is, culture, language, society and history) are in the process of sorting out what is possible and what is not possible before the turn of the century. Some of the trends spotted are peculiar to North America (professional football) and others become icons of memory for America, Europe and East Asia (contemporary Christmas). However, iconoclasts in Russia and elsewhere in the East prefer to destroy the West's icons rather than join in the march towards a world market.

Third, managers and others in the professional class must decide whether America alone, America and Europe together, or the WTO/GATT should lead in completing this part of the big assignment. To decide for the WTO means the U-World dominates choices about trade and investment. To decide for the East and the WTO means America faces a difficult swim upstream and the possibility (but not the probability) of getting screwed in the end.

Wisdom: decision theory lays out possible choices. Pay attention: unilateral choice by either America or Europe for the East leads to Salmon Day, but their joint effort through the WTO might avoid Salmon Day. Open the doors: America and Europe make a choice to defend the West's common interests in the East.

What do I get?

Here is our to do list:

- Buy into the limited role of America and Europe in the East.
- Accept limited commitments to the Orthodox and Moslem lands.
- Build on memories different than those of the West.
- Reimage ideas to include Orthodox and Moslem themes.
- Recognize local elites as 21st century citizens.
- Create a stream of business for North Atlantic Free Trade.
- Use decision theory to weigh costs and benefits of options.

Remember these words: Salmon Day may not be inevitable for the Orthodox and Moslem communities of the East. However, Salmon Day is more probable in these conservative areas than in Mexico and southern South America, East and South East Asia, and in Europe and North America.

Successful decisions

A successful America makes the following policy decisions:

- America works with Europe to create a North Atlantic Free Trade Area.
- America works with East Asia to create an Asia-Pacific Free Trade Area.
- America and Europe let the WTO help Orthodox and Moslem communities in the East.
- America gives limited assistance to Mexico within NAFTA.

Part 2

Europe Finds the Way

How to reimage Europe's place in the global economy

Businessmen's Europe
Euro-Realism
The East

Executive decisions

Remember Europe has choices. In Chapters 4 through 6, the following choices are discussed. They are divided into two categories. The first set has a high probability of success in avoiding Salmon Day for Europe. The second set has a low probability of success in dodging Salmon Day for Europe. You decide what to do.

Decisions leading to success

- The EU uses a two-step approach in setting up the EMU, ECB and the Euro.
- The EU employs a multistep approach in admitting new members from the East.
- The EU permits nation-states to opt out of all new integration efforts.
- The EU begins work on a North Atlantic Free Trade Area with NAFTA.

Decisions leading to failure

- The EU forces all members into the EMU, ECB and the Euro.
- The EU and west Germany subsidize east Germany indefinitely.
- The EU subsidizes three Central European countries for a long period of time.
- The EU supports the accession of all countries in east Europe.

The work of Salmon Day is about making decisions under uncertainty.

Chapter 4

Businessmen's Europe

Europe misses 'the anger, the passion, and the partnership that constitutes the lifeblood of politics.' It makes for a 'businessmen's Europe,' not a 'citizen's Europe'.[1]

Shirley Williams, Britain's European Commissioner in Brussels.

'I am representative of what France is becoming.[2]

Ernest-Antoine Seillière, the managing director of Compagnie Générale d'Industrie et de Participations (CGIP).

Memories always win

'So, are we or aren't we?'[3] boast business executives from Frankfurt, Paris, Amsterdam and Brussels. But what they are remains unstated. It is a catch-all phrase, infinitely adaptable in its ambiguity, expressing all the heteronomy, the subjection to something else of these European Germans and French, these global, largely American-style managers who promise this time to beat back the American challenge. Selling Fords or VWs in London and Hamburg, or distributing Coke or Pilsner beer across the continent, or peddling Mexican–American tacos or Italian pasta to Europe's international elite, these European executives come out on top – or hope to come out on top. Theirs is a college of memories – the story of US domination of the West since the end of World War II.

Here is today's conversion story. Trend spotters see Europeans as the new disciples of *gringo* food, especially Americanized Mexican food, the salsa (loaded with diced tomatoes), picante sauce (a jalapeño pepper and onion mixture used as a condiment on everything from eggs to enchiladas) and chicken fajita (made with cilantro).[4] According to Bob Messenger, editor of *Food Processing*, an industry publication in Chicago, 'whatever becomes popular in the American culture forces itself on the rest of the world'.[5] Pillsbury, a subsidiary of London-based Grand Metropolitan plc, sells tortillas, tacos, refried beans, and mild 'cool salsa' (the mildest form possible) in Britain and Germany under the brand name Old El Paso. The *gringo*-ization of Mexican food continues in America and it is well on its way to capturing the stomachs of northern Europeans, too.

Here is tomorrow's future. Trend spotters see the Finns as the new devotees of the wired world. Banking, shopping and socializing are migrating from the real world into the virtual world faster in Finland than anywhere else, save Silicon Valley in California. 'Finns use the Internet to chat about their problems and for academic research. They use "electronic wallets" and cash systems to order stamps, pick lottery tickets, bank, play roulette in an electronic arcade, shop for groceries, read newspapers and attend courses given hundreds of miles away.'[6] The five Nordic countries (Finland, Iceland, Norway, Sweden and Denmark), the US, Australia, New Zealand, Canada, Switzerland, Singapore, the Netherlands, Denmark, Britain and Austria are among the top 15 in having proportionately the most Internet-connected computers.[7]

'Being wired all the time is the shape of the future', says Johan Helsingus, chief executive of Eunet Finland, for Europe, for America and for the world. 'Of course, America has an electronic culture that is very advanced but it is concentrated in small groups of the population. Here, [in Finland, the other Nordic countries and elsewhere in northern Europe] it is really widespread,'[8] says Minna Puirava, a project manager of marketing studies at the University of Tampere, about 100 miles north of Helsinki.

In short, most 'Americans' in Europe, as well as the Europeans who have to work with them, are wired all the time. The Americanization of the wired culture continues in northern Europe. Also most 'Americans' in Europe, as well as the Europeans who have to work with them have a hamburger, French fries and a coke at McDonald's, eat American-style pizza at Shakey's, buy a bucket of chicken at KFC and munch on donuts and coffee at Dunkin' Donuts. The *gringo*-ization of fast food continues in western Europe.

Of course, most 'Americans' in Europe, as well as the Europeans who have to work with them, speak English and at least some French or German, not to mention a little Italian or Spanish, but no Finnish. It is never exactly clear who they are or aren't, but to all executives, including those who don't speak English, this wordless phrase still captures the hectoring self-confidence of Europe's Americanized business rulers. 'Are we world class executives or aren't we?' Conversations are littered with words from management consulting, like chaos theory, competitive strategy, marketing management, short-cycle time, price-driven costs – the *lingua franca* of boardrooms and offices from Los Angeles to Berlin, in spite of the continents, oceans and nation-states that divide business executives from their common destiny.

Is time running out for the global, largely American-type executives? Should they adapt themselves to a continental, largely German approach to business management whose eyes desire the security of resources from east Europe over the potential of further riches from west Europe? Can this ascendant high-tech international elite preside over a global economy punctuated by ethnic and communal conflicts of extraordinary brutality? Are these managers feeling the pain of longing for a North Atlantic community they never knew they wanted until this cosmopolitan, multicultural free trade world looks like it could slip away? Will these Americanized loyalists continue to lead the relentless march of the global economy through their nation-states?

Economic memory

For example, today's French executives face spiraling unemployment, slow growth and social tensions as Europe prepares to launch a single currency – the European Union's solution to global competitiveness. They are believers in free trade solutions, then, aren't they? They are believers in*pouvoir politique* – or the imposition of political control over Europe's monetary policy, the European Central Bank and the Euro – then, aren't they? Pay attention to the ongoing problems.

First, France, Germany's crucial partner in creating the Euro, is lurching back to serious industrial strife by undoing the welfare state and the very foundations of its national society; when government tampers with what the French call their *acquis* – the established cocoon of a protective state – domestic reform is stillborn because it threatens the core values of the nation.[9] Of course, French solidarity with Germany over the future of the EU is in question.

Second, direct investment in Europe, North America and the rest of the world may be creating a permanent and expanding surplus in manufacturing output that cannot be absorbed in the global marketplace. The French have great difficulty with what they now call *la pensée unique* – that is, the American-rooted ideology that a largely untrammeled market economy is good for the French, for other Europeans and for mankind in general.[10]

Third, labor-saving technology, lower labor costs and higher productivity in East Asia might even be taking jobs away from workers in France, the Low Countries and northern Italy. The French worry that they do not know what Europe to expect, what North Atlantic community to desire and what East Asian tie to dream about.

Therefore, how can Europeans defend the West's industrial structure and employment within a regime of free trade for the lands between California and the German–Polish border? Are European executives ready to be equal partners in the global, largely American economy, pushing free trade across the North Atlantic and to the Pacific, to the East and to Latin America? 'So, are they or aren't they' international decision makers?

U-World versus loyalists

Here is what we know so far from Part 1. The cultural facts between English-speaking Canada and the US are different in detail. Business executives have a common-sense grasp of the similarities and minute differences between football played within the CFL and the NFL. They give them meanings that are more important to Canadians than Americans. These shades of gray play out in Canadian TV and magazine ads for Tim Horton's, Canadian Tire and Petro-Canada.

The social facts between North America and Mexico are different in substance. In this case, managers intuitively have a grasp of the radical differences between football and soccer. These executives give national sports metaphors discontinuous meanings. Nevertheless, the difficult cultural facts are crucial to understanding power in the marketplace, to portraying conventions in the eyes of cosmopolitans and loyalists, and to building on the subjective forces inherent in selling and marketing goods successfully at home and abroad.

Here is where we are going in Part 2. Europeans are born in a society in which they recognize substantive cultural differences. They have internalized the radical divergences among and within nation-states, fought wars over ideological exploitation and tried to impose a semblance of order through such multilateral devices as the European Union. A generation of Europeans has grown up with internal free trade within west Europe as the universal learning device, equally welcoming to French, German, Italian, British, Spanish and Swedish national cultures. So understood, the EU imposes no greater constraints on the contents of Polish, Czech, Hungarian and Slovene national cultures than it did on Irish, Portuguese and Greek contributions to Europe and the West. Preferred national cultural traits are those that make an on going contribution to building a more integrated Europe.

Win–win world

The U-World is at the end of the beginning of a new win–win world market era for breweries. American-owned Budweiser is a niche beer for socializing after work in Japan, and a mass market for watching CFL, NFL and FIFA sporting events on TV in Canada, the US, Mexico and Europe. Budweiser is a global brand. Canadian-owned Molson, Mexican-owned Corona and Australian-owned Fosters are niche beers for free spending young urban professionals (yuppies) in the US. And Belgian-owned Interbrew, the sixth biggest brewer worldwide by volume, is pushing forward its 200 specialty and regional beers (for example, American Rolling Rock, Canadian Labatt, Belgian Hoegaarden, Mexican Dos Equis, Chinese Blue Sword) everywhere it can in the world.[11]

The problem for Interbrew is to be sure the world likes the taste of its regional and specialty beers. Many of its traditional Belgian beers, such as Hoegaarden and Stella Artois, are a bit bitter and an acquired taste for most Americans, Canadians and Japanese. The firm's difficulty is to build on local markets, local brands and local management with local vision. However, all professional managers remain at the beck and call of three powerful family clans who hold all the company's shares, make strategic decisions and review financial performance. The family wants its executives to use more of the firm's portfolio of 49 richly flavored Belgian brews in their response to market pressures from all over the world.

'What needs to be done?' Find more markets for Belgian and acquired foreign beers. 'What do we get?' A niche strategy multiplied tenfold all over the world. 'What can we do?' Use strong cash flow to make another acquisition the size of Labatt (about $2 billion in US dollars). 'Tell me what you are going to do on Monday that's different?' Look more carefully at local market information about mass market versus niche consumption, and make some Belgian beers less bitter for the global, largely American taste.

Both Budweiser and Interbrew are practising good intellectual asset management. Budweiser reimaged its beer into a global brand, and Interbrew reinterpreted itself into a brewer of regional and specialty niche brands. Both grew sales all over the world. They have a win–win world. On the other hand, neither Molson, Corona nor Fosters ever got beyond capturing a small niche in a very big US market. These three firms failed to practice good intellectual asset management, and remain also rans in the race for market share. They face a win–win–lose world and are candidates for Salmon Day.

Paying attention to free trade

Today in Europe we have the autonomous cultural dynamics of a businessmen's Europe. They are generating copies of themselves with a recent mutation in which the German model is evolving into a stronger position *vis à vis* other European managerial habits, and the American set of cultural traits is dominating all contenders within the North Atlantic community. Europeans are in the process of selecting those managerial habits that suit them the best, and some British, Dutch and Swedish are propagating those from the US rather than adopting others from Germany. The cultural evolution of the cosmopolitan elite within Europe is a positive force for economic growth on both sides of the North Atlantic. These international managers have grown familiar with free trade under the EU and NAFTA, and they see the approach of new members as increasing the chances of survival in the management of multinational firms and the latter's interlocking relationships with nation-states throughout the world.

Enthusiasm for broadening the reach of the European Union could be a distraction because of the energy wasted on trying to integrate Polish agriculture, to mute the ethnic rivalries of the Balkans and to end the backwardness of infrastructure in the former Soviet Union. Perhaps the cosmopolitans should put their attention to resolving some of the problems in east Europe before trying to tailor the admission of these new nation-states to the rigidities of a Brussels-led EU. It is a mistake to try to overcome all these ancient and recent mistakes in economic choices by signing up most central and east European countries as full members of the EU.

Free trade is an evolutionary concept. It welcomes particular national cultures at different speeds, and it plays some role in bringing stability to poorer regions as they seek their place in the global economy. Even with hundreds of years of experience in free trade the American South and the Canadian Maritime provinces are still vastly poorer than the rest of the US and Canada. Also even with decades of experience in free trade the Italian south and the west of Ireland are poorer than the other parts of their countries and of Europe. Moreover, even with less than a decade of experience in free trade, east Germany (the former Communist part, or DDR) is still way behind west Germany in terms of productivity, output, jobs and wealth.

Free trade is an idea that has spread and stabilized throughout the West and parts of East Asia. Free trade transforms itself each and every time it is transmitted – from the death of the Mexican middle class after the peso devaluation, to the protectionist common market called Mercosur and to the belief in a common currency for solving all of Europe's competitive needs.

Free trade is a resilient idea. Free trade is a contagious idea. Free trade has the power to inspire others to join in the quest for wealth-creating prosperity.

Rooted cosmopolitans and loyalists have organized the thought processes of those who transmit ideas to raise up free trade over all its competitors. Free trade is a universal concept; it has an immutable underlying essence – that is, free movement of goods (no tariffs) among member countries. Free trade in North America reflects the need for free mobility of capital from the US to Canada and Mexico. Free trade in Europe reveals the preference for open borders, and the need for Europeans of all nationalities to get to know one another so the post-World War II era is simply the end of the beginning of a prosperous twenty-first century. Free trade in Asia suggests the intent of linking up loyalists in Japan, China and India for a better competitive position within the global marketplace. In short, free trade varies from culture to culture, but it varies in directions that are predicted by preferences for economic growth and more open markets worldwide.

Learning to pay attention to free trade within the cultural memories of Americans, Europeans and East Asians carries with it the need to understand the societal, language and historical meanings of free trade. These are discontinuous cultural facts, which are biased as they move along their evolutionary path and are transformed into an organizational regime called NAFTA, EU, Mercosur and APEC. These multiple efforts at free trade have not stabilized in the same manner. When new members are added or new programs for integration are tried, they will restabilize in unforeseen ways among nation-states, between loyalists and cosmopolitans, and with emerging economic forces outside the free trade area. Let's look at the proposals to both deepen and broaden the European Union, and their impact on whether Europe can avoid Salmon Day.

Free trade in crisis

Within North America a good bit of the discussion about NAFTA comes from the deductive knowing–believing paradigm, in which commentary begins with the assertion of a theory with potentially universal application. Since free trade worked well for the 50 states of the American union, it could work equally well for the ten provinces of Canadian confederation. Both are federal nation-states in which sovereignty is shared with the states and provinces. Their joining together across the North American continent brings Quebec–the Maritimes–New England–the Mid-Atlantic States, Ontario–the American Midwest, the Prairie Provinces–the American mountain states, and British Columbia–the American northwest–Alaska into a redeployment of assets on both sides of the open border. No need to do a great deal of empirical investigation because ten years of transition is sufficient time for all industries, firms and managers to realign their manufacturing and marketing to take advantage of the new home market realities in North America.

This knowing–believing paradigm was used to transfer the religious belief in free trade to Mexico. Since free trade worked well for the combined American–Canadian effort, it would work equally as well for the unitary state of Mexico. Firms would redeploy their assets within Texas–northern Mexico or California–Baja California and both sides of the frontier might prosper. Yet the border industry program in low value-added manufacturing took place on the Mexican side of the frontier and did well in terms of jobs created in Mexico, but it had been going on long before continental free trade became a goal of the Mexican government. However, central Mexico (where the bulk of Mexico's population lives) could not absorb the flow of American goods into the country, and the country eventually went bankrupt.

Now some want to extend NAFTA to other countries in Latin America and to broaden the EU to the eastern countries in Europe. Proponents suggest free trade has universal application without raising important questions about its specific cultural and historical context. Historians call these questions *problématiques*, and this approach comes from the French *Annales* school. Let's see whether we can adduce answers to specific questions (that is, let's emphasize the particular) so that we have a satisfactory explanatory fit to the logic of free trade for Central and east Europe.

Europe and America

'Are we Europeans or aren't we?' The Germans at BMW brag after beating the Japanese in the Rover acquisition. The Swedes at Electrolux swagger after regaining the largest market share in white-line appliances from Whirlpool. The French–Germans–British–Spanish croon after pushing Airbus into a strong number two position behind Boeing. It means: 'Did we show the Americans or didn't we?' Today's America is a foreign country to European business executives. It is aggressively modern, ceaselessly uprooting the past, an America whose mind's eye wants to capture the cultural facts from East Asia and Latin America. Today's Europe is the Old World to American business executives. It is more traditional, carefully protecting the past, a Europe whose mind's eye seeks to capture the cultural facts of the more conservative Orthodox community to the East. Here is the unhappy truth for Europeans: America's power over Europe within the North Atlantic community comes through capturing insights from East Asia in divining the global economy.

Free trade is becoming an obsession because of what it means to those inside and others in 'the near abroad.' How imperfect are all decisions about which countries to include and exclude, how questionable are rigorous admissions criteria, how utterly beyond definitive interpretation are successes and failures. Any discussion of free trade ends up as a series of anecdotes about how to calculate value added, to do substantial transformation, to overcome customs difficulties. Confessions are made as to why Greece is in and Turkey is out of the EU, and *sotto voce* why Hungary will not get in without Romania, too. *Te Deums* are sung as Germany sacrifices France and the rest of the EU to the demands of the new currency, the Euro. Europeans feel more comfortable with trade certainty and monetary ambiguity than with the titanic struggles in Brussels to create a European nation-state out of the tattered cloth of many ancient national cultures. In this, they are traditional Europeans, not aggressively modern Americans.

All of this goes by the name of globalization. For Europeans, globalization is a euphemism for American hegemony; for the pervasiveness of American pop culture and fast foods, the power of American capital markets and companies. Globalization has become the standard polysyllabic description of what shapes the world as the century ends. What if Europeans don't want it? What happens when the economic realities of globalization force European workers to give up habits and benefits to which they think they are entitled? Or if the coming of the global, largely American culture is perceived as threatening the European peoples' sense of identity? Note: a backlash against globalization means growing anti-Americanism: the two are the same thing.

How does this potential reaction to globalization affect free trade? Perhaps the European Union's march eastwards towards the American ideal of free trade is not irreversible. Perhaps 1998 is the make or break year for Mexico, Latin America, and Central and east Europe.

Europe, America's frontier for trade

Is this hesitancy in America's free trade trek to the south and in Europe's free trade procession to the east all that worrisome? Yesterday's rush pell-mell by the twin pillars of the North Atlantic community to drop everything worth having in the world of trade and investment for the slim chance of development success in the adjacent poorer lands looks unseemly. Today, both America and Europe have more important things to do to grow their economies and to create wealth for their people.

Here are the particulars. Europe remains America's most important economic partner and will be for years to come – are we or aren't we? 'The Economic Strategy Institute, a private research group financed by American corporations, compared Asia and western Europe in categories like trade, foreign direct investment in the United States, American direct investment abroad and technology transfers. The study argues that America has much more balanced trade with Europe, and more mutually beneficial commerce, because the two markets have basically similar models of capitalism [with a normal distribution of the flows of goods, capital and technology].The study concluded that the links between the United States and Europe benefited the American economy far more than those between the United States and Asia.'[12]

After years of productivity gains in the US, American firms are going into Europe to exploit the new markets opening up to all comers from around the world. They prize the EU's prosperous middle-class markets over Asia's growing economies, and they want to provide their new customers with American-style service premiums at lower price points. American firms are bringing these New World values in finance, retailing and technology to Europe and European firms are fighting back with their own competitive strategies for growing sales and increasing profits. America is definitely shaking up European business practices. That is the news story.[13]

Let us try to adduce answers to specific questions and see whether we have an explanatory fit between what is actual and what is fantasy.[14]

How many converts practice the American way of doing business? Some from Britain and the Netherlands, a few from Germany and still fewer from the rest of the EU.

How many European companies list their shares on the New York Stock Exchange? More every day because listing opens up new sources of capital. Hence the large number of firms from Britain, Germany, Italy and the Netherlands, and the fewer number of firms from Finland, Sweden and Portugal that trade on the NYSE.

How many European firms use American accounting standards? Not that many because Britain has its own generally accepted accounting practices, and the rest of the EU is bound by practices emanating from a different legal tradition.

How many European enterprises link executive bonuses to the price of shares? A few, but this is not common practice because European executives pursue a long-term industrial strategy that encourages high operating profits instead of returns to shareholders.

How many European executives aim to grow shareholder value by increasing returns on equity? More today than yesterday because this is an idea whose time has come for business executives on both sides of the Atlantic. However, European firms still do not make good use of their capital. For example, between 1974 and 1993 in Germany the return on capital was just over 7% compared with 9% in the US. Also the return on equity in German companies in 1994 was only 7.4%, half that in American firms.

American business practices come from the cultural soil of the US. A few may grow in the European earth; most will have to be cross-bred with European business practices. What may bloom could be something stronger, perhaps better able to withstand the storms of controversy among nation-states that seek a place of influence within the EU.

America, Europe's teacher

Are we, or aren't we capitalists who grow our business based on a personal trust between bankers and executives, managers who share power with government, labor and other stakeholders, top people who are drawn from groups that all knew one another as children, university students and professionals. Relationship management is the jargon today in American consulting for this Old World custom of building trust through cross-share holdings among banks, insurers and industrial companies.

Some Americanized French give these relationships a negative spin with the phrase 'statist meccano.' Italians suggest these ties are good for everyone and call them *salotto buono*. Nevertheless, if lenders and shareholders did not go to university with the chairman, they do not get seats on boards of directors. To do otherwise would break down social cohesion, lead to hostile takeovers and other opportunistic behavior, and create an American style of business in Europe.

Clearly, American capitalism is not going to bring about the collapse of civilized business as practised by European executives. What will really happen is as follows: changes will continue among European management as they learn from America, but these will evolve slowly over the course of time. Revolution in managerial relationships is neither in Europe's best interest nor its strong suit.

Europe still has a long way to go before it learns all the lessons America might provide. And that is good because, as Europe learns from America, America learns from Europe. After all, both of us together are the West. Are we or aren't we one or more of the following:[15]

- Inheritors of the Renaissance – a period in which the restless quest for wealth and material possessions was the essential ingredient in the rebirth of European civilization.
- Map makers in the Age of Discovery – a time of wider international exchanges that ignored conventional kingdoms, ancient religions and antiquated beliefs.
- Iconoclasts who during the Reformation split western Europe in two and carried their theological earnestness to its logical conclusion, a di-

vided North and South America; who during industrialization rejected mercantilism's siren song of protectionism for the hopefulness of free trade; and who during the movement towards democracy toppled kings, commissars and military regimes so the people could rule.

- Advertising executives in the postmodern world – a present day filled with exuberant multiculturalism and bravura consumerism, the worldly goods of Santa Claus found to be so repelling and fascinating by devout Roman Catholics, Evangelical Protestants, Eastern Orthodox, Moslems, Hindus, Buddhists and all others who want a taste of the global culture without giving up all their traditional beliefs.

Since the mid-fifteenth century, at the very beginning of the Renaissance and the Age of Discovery, when Asia and America began making Europe, we in the West have been avid materialists. Did we or didn't we want luxury goods from the East Indies – spices, silks, tea and pewter – so much so that merchants and kings in the West acquiesced in the conquest of Constantinople (the spiritual home of the Eastern Orthodox) by the Ottoman Turks in 1453. All we wanted is unabated trade with the East (which we got). Did we or didn't we want oil from the Moslem East so much so that Christian presidents and prime ministers used all the fire power at their disposal to route an army of believers in Allah, Islam's name for the monotheist God. Do we or don't we want goods and services from all over the world – cars, computers, satellite TV, Internet connections, mild cool salsa, pasta. The list of worldly goods goes on without ending for we always ask for more. Do we or don't we make hardheaded decisions to make money, increase profits and raise returns to shareholders.

What Europe taught America and America is now teaching Europe is how in the West materialism triumphs over spirituality. What the West taught East Asia and East Asia is again teaching the West is how secular materialism prevails over religious excess. What the West wants to teach the Orthodox and Moslem East and the traditional Mesoamerican societies in Latin America is how worldly goods mean more to mankind's life on Earth than candles, prayer rugs and medicine men. After all, the West's culture of worldly goods has been here for over 500 years – are we or aren't we image breakers, prophets and teachers, advocates of this new religion called globalization?

Translation failures

Until the mid-1980s, countless management texts were written exclusively by American gurus about business practices within multinational enterprises. Their difficulties in translating the complex social processes inherent in European industrial policy shows up in the artificial linguistic and cultural boundaries set up by American management experts. Unfortunately, much of what makes Europe gets lost in translation because American consultants rarely have a good grasp on the contexts out of which and into which German or French business decisions are made. American gurus do not have a sensitive feel for the macro structures of movement among national histories, languages, cultures, societies and national economies.

On the other hand, Europeans know that as they shift countries, so they shift perceptions about how to go about building global businesses. Europeans understand without saying anything that they could impute success at Sweden's Electrolux but such discussion might look foolish at Italy's Zanussi. The phrase 'return on equity' is translatable in that national accounting and financial groups give us equivalents in British English, German and French, but the implications of the concept for managing business firms are not transferable across cultures.

What distinguishes the 1990s Euroguru studies from previous American guru studies is that Europeans have a leg up in three important areas.[16]

Cultural diversity

Eurogurus know without saying that attitudes towards pay, the *acquis*, stakeholder value, social obligations and industrial policy differ vastly in Britain, the potato belt of northern Europe and the pasta belt of southern Europe. England perceives itself as different from the rest of Europe and much closer to America than all other countries that supply ideas for use in the New World. France offers itself as a model of civilized habits even though American managers take nothing from the elite clubbiness of French enterprises. Germany cannot help itself as its economic power compels it to extend the trust between bankers and industrial companies to nation-states within the EU and those in Germany's 'near abroad', such as the Czech Republic. Britain, France, Germany and others in Europe are faithful to their own managerial traditions rather than to European or American executive ideas.

'Soft' management

Eurogurus prefer the human side of management–employee relations over the American emphasis on monetary incentives. Many European countries have histories of nationwide labor strife over pay, the workweek and retirement benefits that are individual corporate issues in the US. Britain rejects the dominance of leftist-leaning labor unions and seeks instead to explore American practices as a creative act of modernization. Germany makes labor unions members of supervisory boards of corporations so they will join in solutions rather than create additional difficulties for management. France, Italy and others grapple with the '24-hour strike' and the parked trucks on the superhighways as union syndicates beat back all government attempts to lower costs and raise productivity so the countries can join the new EU currency system, the Euro. Instead of harping on about the differences, we need to look carefully at what a national culture gains, and at the ideological implications of soft management over scientific management.

Global companies

Eurogurus point out the parochial views of their American management counterparts. In this, the former are correct. There *is* very little German, French or European context in American texts, and that has as much to do with management traditions and conventions as it has to do with international economics. Claims of complete ignorance about the British–Dutch Unilever and Royal Dutch Shell, or about the Swedish ABB and Electrolux are simply naive. Many American management gurus recognize the global nature of these European firms, and the popularity among American readers of the lessons taught by these firms for P&G, Exxon, Caterpillar and Whirlpool. The canon on global business includes both sets of firms as well as those from Japan; for example, Kao, Komatsu, Sony and others.

There is a useful reminder in all of this discussion about gurus. Americans, English-speaking Canadians and the British are resolutely monolingual, and they are unaware of other cultures as a consequence. Management books can be used in cross-cultural teaching so long as they are written in English and they don't suffer from translation failure. On the other hand, Italian management texts suffer from a business schizophrenia that distorts both past and present. Centuries of preference for French among northern Italians strengthens the preference for older, more traditional social habits over newer more modern business practices from America. Only in the 1990s have a few Eurogurus moved out of the ghetto of management isolation into the mainstream of EU integration and NAFTA free trade. As long as Eurogurus start with the American lack of interest and a reluctance to theorize, they will be read, studied and used by global, largely American managers.

The work of gurus in shaping management practice does not take place in a vacuum but in a continuum, and knowing more about how it is perceived on both sides of the North Atlantic can assist executives in doing the work of global business. The study of management gurus can teach us a great deal not only about other business cultures, but also about our own.

Therefore, global managers should read Charles Handy, Yves Doz, John Stopford, John Kay, Sumantra Ghoshal, Geert Hofstede, Manfred Kets de Vries, Arie de Geus, Percy Barnevik and Wisse Deker – all Eurogurus – as well as Peter Drucker, Michael Porter, Tom Peters, Gary Hamel and Phil Kotler – all American gurus.

Translation successes

'I am representative of what France is becoming', says Ernest-Antoine Seillière, the managing director of Compagnie Générale d'Industrie et de Participations (CGIP).[17] Its market capitalization is Ffr16.6 billion ($3.1 billion in 1997 US dollars). The holdings of CGIP include Crown Cork & Seal (packaging), Cap Gemini (electronics), Valeo (auto components) and companies in biotechnology and abrasives. At CGIP, 'influence and strategy, the obsessions of many prominent French managers, have been tempered by a concern for financial discipline and profitability ... unrefined Anglo-Saxon [American] values ... [Yet] reform seems to be measured in decades ... [and this] is not quite radical enough.'[18]

During the 1990s, Daimler–Benz, Germany's largest industrial company and its biggest subsidiary, the auto manufacturer Mercedes–Benz, recently took several actions that show German managers are learning a thing or two from their American counterparts.[19] First, through a reorganization it reintegrated Mercedes–Benz into Daimler–Benz. Second, the firm diversified the product line in automobiles to include better value products (E-Class saloon), niche products (M-Class four-wheel-drive vehicle) and small cars (A-Class hatchback). Third, the company internationalized itself with a new, lower-cost plant in Alabama, and lowered its exposure to foreign exchange risk *vis à vis* the US dollar.

Both groups of national managers are learning, each in their own way, from both European and American management gurus. Is there a European managerial style or isn't there?

Germany and France

To the two countries that were to become the translators of Europe to the world, who met for the first time in the 1950s in the European coal and steel community, this was indeed a very large, very strange world. Past their first, angry, tentative declarations spoken with true German forcefulness and French subtleties, something like intense consummation joined their speech and suddenly the two strangers, who had fought three brutal wars to the death without thinking about the consequences, began to build together the Common Market of six nations, the European Community of 12 nation-states and the confederal European Union of 15 member states. Each was accompanied by America, which was neither liked or wanted at the councils of Europe but whose job it was to coach these old enemies (now friends) into a deal. France would not think twice about blackballing

Britain because the former set the political agenda for Europe. Germany would go along because it wanted stronger economic and financial institutions that tied it to the whole of Europe.

Hence the frantic search for regional subsidies, the tireless demands for agricultural supports, the dramatic drive for a single integrated market, followed by epic concerns for Europe's health, waiting for confirmation that the remedies are solving Europe's problems. Wrapped in such integrating macrostructures as the Council of Ministers, the European Commission, The High Court and the European Parliament, the laws, regulations and decisions were so opaque they confused national governments and multinational enterprises alike. The people would find out soon enough and make a fuss. But European integration came first, Brussels told the member states, who, more than ever now, were convinced that this truly was a new sovereign, though Brussels noticed – as others did not – that the center's successes were fewer in number and more difficult to achieve.

Maastricht arrived, stained by the opt-out provisions of Britain, an almost negative vote in Mitterand's France, a reluctant second referendum vote in Denmark, and a dithering European Parliament's unwillingness to add new members quickly and pass the budget without dotting all the 'i's and crossing all the 't's.

That was the cue, for on hearing of Maastricht's difficult birth, France and Germany lunged towards political and monetary union for Europe, while the other members struggled with displeasure and annoyance, like countries being occupied during war, feigning outward respect and seething internally – what are we to do, are we or aren't we equal members, too. But then, seeing that Germany and France refused to release their hold on the levers of power, the other members would slacken and cease to fight and make suggestions about perfecting Europe's new central bank, the monetary institute and the Euro, sitting at the Council of Ministers as if to make out whether Germany would really give up its beloved D-mark, finally taking a deep breath, filled with premonitions about deep budget cuts and increased taxes, and the yearning to get it over with long before 1999. All France and Germany had to do was squeeze a bit harder, and out would come Italy's '*Si signore*' that Rome had hoped never to say.

Great Britain

Whenever the two partners met, they never failed to hint about Britain, to let it be known that England's needs were their most important European problem – all of these efforts mysteriously subtle but unnecessary, as it would never occur to the French and the Germans that Britain was really crucial to the forging of an integrated Europe from the Atlantic to the Vistula. Is Britain European or isn't it?

The late 1990s are the choicest years for the partners and they never let Britain (or any of the other members) interfere with them – not slower economic growth, increased unemployment and the higher barriers to keep America out of Europe. At this time, Germany and France struck their bargain. Germany got political union so it could play a commanding role in the foreign policy of a united Europe. France got monetary union so it could regain some control over its financial condition from Germany. And the whole of Europe got the prospect that it will be free from another German war in the twenty-first century.[20]

Italy and Spain

However, Italy's conversation about the partners is mostly a repertory of unflagging complaints:[21]

1. Germans believe Italy is the bearer of instability and inflation.
2. Italians believe Germany is throwing its weight around.
3. Italy screams at the horror of the straightjacket in which the Deutschmark and the Bundesbank have new names, the Euro and the European Central Bank.
4. Italy shudders at the technocratic conception of Europe by France whose own history is Colbertism and whose modern equivalent is Brussels.
5. Both Italy and Spain know that they do not meet the strict Maastricht criteria for joining the European monetary union, and they fear France will give in to Germany and keep both countries out.

Is what Germany and France want good for the rest of Europe? Is France's monetary subordination to Germany good for Europe?

Since the partners were always precise in their comments about the Club Med or Southern Comfort countries, once the mood for negative feelings had set in, France and Germany would refer to Portugal, Italy, Greece and Spain as the PIGS.[22] If they get into the Euro, it will make the new currency weak in terms of the US dollar. Europe's exports will be priced competitively, but the returns to investors on European stocks and bonds will fall. If Club Med is let into the Euro, the partners will hurt their economies; if the PIGS are left out, the partners will get hurt feelings and, perhaps, create a split in the EU.

In or out, a decision about Italy and Spain is a political one for Germany and France, but a 'yes' decision, one that includes all members of the EU who want to give up their national currencies for the Euro, is a crucial one for executives and managers. They want a businessmen's Europe, one with a single market, that is single-minded, and has a single supra-national currency.

Europe as a rooted cosmopolitan community

With a few short years left before the Euro becomes a reality for Germany and France – and for much of northern and southern Europe as well – Brussels has lost its appetite for new legislation to complete Europe's single market. Here are some examples of obstacles to free trade.[23]

- Gone is the urge to harmonize tax rates or tackle company law, which are two of the biggest obstacles still unresolved and which affect the single market integration of the financial services and insurance sectors.
- Gone too is the thought that Brussels should enforce all existing legislation. Instead, the nation-state members need to accept more of the certification and test results of other member states rather than impose national solutions to truly European problems; for example, Germany

requires packaging materials to be recycled, and this plays havoc with the ability of French exporters to sell products across the Rhine.

- Gone also is the ability to offset profits in one country against losses in another member state, and pay taxes only on net EU profits.
- Gone is the hope that people could move around Europe without passport controls.

The Confederation of European Industry hopes the single currency, the Euro, will make the market more transparent in terms of prices and costs for business executives and consumers. Europe's international elite see the EU as the means for making the whole of Europe more cosmopolitan and less a creature of loyalist nation-states.

North and south, east and west

Up to this point, the north–south income divide in Europe is closing as regional aid from Brussels reduces social and economic disparities in the EU.

> *'Ireland has made the most spectacular advance among the [poorest EU countries] in raising incomes through higher economic growth. Thanks to annual growth well in excess of 5 per cent, Ireland's gross domestic product per head has risen from 63.6 per cent of the EU average to 89.9 per cent in 1995.*
>
> *'Spain has moved up form 70.5 per cent in 1983 to 76.2 per cent in 1995. Portugal has climbed from 55.1 per cent to 68.4 per cent. Greece has only raised its income per head from 61.9 to 64.3 per cent, despite receiving hundreds of millions of ecus of aid.*
>
> *'Among the regions, Hamburg retains its top spot as the Union's wealthiest with 189 per cent of average EU GDP per head. Brussels (183) is second, swapping places with Ile de France around Paris (163); Greater London (144) to ninth.*

> *'The poorest regions include Saxony (53) in former East Germany, Galicia (60) and Andalucia (58) in Spain and Calabria (61) in southern Italy.'*[24]

Only Saxony (and its important industrial cities of Leipzig and Dresden) has the opportunity to pull itself out of the category of a poor region as Poland and the Czech Republic become full members of the EU. Both Central European countries lie next to Saxony and they will add to the latter's average GDP per head.

Expansion to the east means more aid to Saxony, Berlin–Brandenburg, Poland-A (the former Prussian-German land), and Bohemia and Moravia, and less aid to Greece, southern Italy and southern Spain, Portugal and Ireland. Will the existing member states permit the addition of new members in the east with the knowledge that the central and east European countries will need a lot more aid? This is still an open question.

The grand purpose of Europe

Only now would it hit the rooted cosmopolitans of Europe, this truth about the EU, that free mobility of people, the European Monetary Union, the Euro and the expansion to the east form the grand purpose of Europe. Without these events, the EU will not be able to define the identity of the peoples who live within the 15 member states and those who live within the countries that want to be members of the EU. Only now would it hit the loyalists of Europe, this truth about the EU, that without these four utilitarian actions – all difficult, all necessary – Europe will become what it once was, a quarrelsome group of nationalities.

The EU turns the lights on everywhere as power shifts to the international elite in Brussels, at the nation-state level, among nongovernment organizations and within multinational enterprises. The EU offers Europe as a whole power over its joint destiny with America. The EU offers its like-minded people a shared trip to the future, what ever that might be.

Europe's big assignment

Europe's hesitancy to carry out big assignments is likely to continue because of the continent's history of multiple languages, cultures and societies. The collective memory of Europe is one of many lands, some with small egos (the Benelux countries) and others with larger egos (especially Germany and France). Together, they have the following remembrances: Europe was once a land of some very big countries that did not like too much change, too quickly. Even today Europe is not yet a nation and, of course, cannot take on the really big assignments.

Yet Europe is on the verge of something to something else. Europe could deepen itself with monetary union, or Europe could broaden itself with expansion to the east, or Europe could do both. The decisive factor in postmodern history is that Europe has a choice. Will it look back or will it go forward?

America hopes Europe will choose a future that includes a deeper and broader union on the continent, and a future Europe as an equal partner with America. The US does not understand Europe's legacies from Europe's past.

First, Germany's three strikes at trying to dominate the continent through war, and its two recent losses.

Second, the rise and fall of the overseas imperial empires of Britain, France, Spain and Portugal.

Third, the inability to integrate the conservative Orthodox communities, especially the Eurasian empire of Russia, into west, central and east Europe.

Lastly, the lasting imprint of tensions in the Balkans from the now defunct Ottoman empire.

Europe is not America and it never will be. Europe is a different place, with a different history and a different collection of peoples, cultures and societies. Europe's icons of memory may look similar to those in America. This is an illusion. Europe's real cultural differences are there in the land and on the faces of its people. America's multicultural facade is nothing compared to the many cultural edifices built up by Europe during the last 500 years.

Costs and benefits

How should Europe weigh the costs of attempting to create a deeper and broader European Union against the benefits of less political and economic conflict in the future?

This is a case of decision-making under uncertainty. If all the European countries involved joined in the effort in a spare-no-expense way (by giving up the status of nation-states), the possibility of further conflict on the EU portion of the European continent would be ended. If the European countries involved joined in the effort on a slower basis, which seems to be the case with the Euro and the expansion to the east, a greater possibility of political conflict would remain. How much greater is unknown.

Decision theory cannot create certainty where none exists, but it offers a way to organize one's thoughts and gather the information needed to make better decisions. Consider economic union. The best policy is for the member states to use the EU to tie Europe together for all those that wish to join.

Remember Europe has choices. One option is to deepen the EU among the 15 member states. Germany and France would lead the effort and share all the costs and benefits. The Benelux and Nordic countries, and Italy and Spain would follow close behind with lower costs and similar benefits. The rest would participate on the basis of 'variable geometry', in which benefits accrue based on costs incurred.

A second option is to broaden the EU among the nations to East. The EU would expand several times until all who wanted to join and all who were qualified to join would be included in Europe's programs of free trade in manufactured goods, a customs union for all goods and services and a single currency for rich and poor alike. The original members from the West would have to subsidize the new members from the East.

Today, the EU is pursuing both options one and two without regard to the downside costs of each, and the lack of joint and several benefits from these efforts. Will Europe prosper from both of these simultaneously? Probably not. Europe might have to make a choice between its ideas about deepening from Maastricht and its desires for broadening to secure the German border in the East.

Option three is to join with America in a North Atlantic Free Trade Area. The thought is to make Europe an equal partner of America. However, this will not be possible unless Europe both deepens and broadens the EU. Thus both Europe and America have a great deal riding on the success of Europe's recent efforts at building up the EU.

Decision theory offers us a way to reflect on our choices and make the best decisions we can with the information we have in front of us. Decision theory offers Europe a way to organize itself to beat the fate of Salmon Day.

Salmon Day

Salmon Day is the experience of spending an entire day swimming upstream only to get screwed in the end. Budweiser, Interbrew and Daimler–Benz fought the global fight, gained market share and won support on Wall Street. None of these firms were victims of Salmon Day, and they represent American and European successes in the late 1990s.

Here are some summary thoughts. First, the transmitters of Europe's common knowledge about the big idea called the European Union offer the following: participation in the global, largely American life style with the chance that some of it might become a European life style as well; English, French and German as languages of communication with the first having the primacy of use throughout Europe and the other two having the primacy of use in their own specific spheres of influence; and Europe's ongoing big assignment – namely, the creation of a free trade between the Atlantic Ocean and the borderlands to the East.

Second, the editors of Europe's common knowledge are in the process of reimaging their intellectual thoughts, conclusions and policy prescriptions concerning the U-World and the world of European loyalist nation-states. Many of their vivid memories of translation failures will remain strong while others will be remembered as problems of transition – for example, from multiple national currencies to one single currency – the costs and benefits of doing a better job of European integration will be spread unevenly throughout European society.

Third, managers and others in the professional class must decide whether Europe expands free trade and how it joins itself with America in completing the biggest assignment of all in the twenty-first century. To decide for America means Europe continues to swim the North Atlantic ocean without serious risk to its political and economic fate. However, to decide for America means Europe joins Canada in swimming the ocean with a serious challenge to its cultural independence. Already, the global, largely American life style permeates many consumer and some managerial habits within western Europe. These changes in cultural practices are bound to multiple when and if Europe and America come together in a North Atlantic Free Trade Area. Here then is Europe's danger as it swims upstream with the probability of getting screwed in the end.

All agree the nation-states of Europe have been strong political nationalizers, especially, the Franco-German partnership that dominates crucial decisions about the future of the EU. On the other hand, the European Common Market–Community–Union is an important emerging economic globalizer that uses European 'nationalism' to strengthen its global position within central and eastern Europe. Budweiser, Interbrew and Daimler–Benz became successful within the EU and in eastern Europe, and in North America and East Asia through the market-rationale approach towards the global economy. Their cross-border and trans-North Atlantic strategies may be signposts to the future for European firms.

Wisdom: decision theory lays out possible choices. Pay attention: one choice leads to Salmon Day, and the others lead to Europe sharing in a second American century. Open the doors: Europe makes a choice to defend its most important cultural interests without sacrificing its political and economic stability.

What do I get?

Here is our to do list:

- Buy into Europe's big assignment for the EU.
- Accept loyalist nation-states as crucial to Europe.

- Build on the memories of national social cohesion.
- Reimage ideas about a common currency and industrial policy.
- Recognize rooted cosmopolitans as twenty-first century Europeans.
- Create a pan-North America–Europe stream of business.
- Use decision theory to weigh costs and benefits of options.

Remember these words: Salmon Day is not inevitable for Europe, but the EU will have to work a great deal harder to avoid Salmon Day.

Chapter 5

Euro-Realism

'If there was ever a bad idea, EMU is it.'[1]

Rudi Dornbusch, the Ford Professor of Economics and International Management at the Massachusetts Institute of Technology.

'It may be better not to have a common currency at all because mounting popular discontent would likely sweep away present policies, including the single currency.'[2]

George Soros.

Money always wins

Managing global business is about judging the certainty of tomorrow's European Monetary Union (EMU), its new European central bank (ECB) and its forthcoming cross-border non-national currency, the Euro. Is this realism or fantasy? Here is the ongoing debate.

Euro-terms

'The Maastricht Treaty ... [defines] the statutes of the future European central bank, which will be politically independent and charged with the principal objective of price stability; the two features characterizing [Germany's] Bundesbank. Thus the Maastricht Treaty has set the condition for allowing the individual countries to choose freely whether they will join or not. By accepting to join they also accept the institutional constraints surrounding the future European central bank,'[3] says Paul De Grauwe of the University of Leuven and a member of the Belgian Senate.

Euro-governance

Once in the EMU, governments gain a voice in setting rules concerning price and currency stability, fiscal and monetary restraint and noninflationary economic growth. All the nation-states that join the EMU become members of the board of ECB. Without limiting the independence of the ECB, they jointly manage the foreign exchange (forex) value of the Euro. This seems to be an oxymoron because some members of the board plan to use their political power – *pouvoir politique* – to assert national control over the independence of the ECB.[4] Such efforts will be resisted strongly by Germany whose Bundesbank is truly independent of the national German government.

Euro-practicality

The Euro will compete against the older national currencies, especially the American dollar and the Japanese yen. Today, we do not know the consequences of the future collective actions of the ECB board. However, we can make one good prediction: the German members of the Board will push the evolution of the new central bank towards the known success of their Bundesbank. For the most part, the other members will accept this direction as the cost for price and currency stability within the EU.[5] However, during times of high unemployment, all members will have trouble enforcing their new monetary rules because Europe lacks flexible wages and functioning labor markets. Labor across western Europe will strike before it gives up wages, the *acquis* and other benefits.

Euro-foolishness

The Maastricht Treaty instructs those countries that do meet the conditions to move to the EMU willy nilly in 1999. At that time, the conversion rates of the national currencies are fixed, and the Euro exists in the computer programs of banks.[6] Will the Euro survive outside, asymmetric shocks, such as a big rise in oil prices, or substantial job loss due to recession?[7] Germany, Austria and the three Benelux countries could work together with fiscal transfers through the European Commission to sustain themselves during these shocks. Maybe France might be supported too by these other countries, but Article 104b of the Maastricht Treaty 'specifically rules out a financial rescue by one country of another'.[8] Thus France might be on its own to reduce public-sector jobs, and stay afloat as a nation-state in a time of great labor strife.

Euro-results

According to Hanns Glatz, head of European affairs for Daimler–Benz in Brussels, 'Currency fluctuations are far greater concerns to us than tariffs. We really don't worry about existing customs duties. [For example, the average tariff between the US and Europe is a meager 3%.] What we really are interested in are identical regulations'[9] for the circulation of bank notes, maintenance of account balances, use of electronic money and all other financial transactions within the EU.

Euro-fantasy

Does monetary union do this for Europe? Here is De Grauwe's answer in 1994: 'As of today ..., it has not been shown in a conclusive way that a monetary union with [15] members makes sense from an economic point of view'.[10] 'If there was ever a bad idea, EMU is it',[11] says Rudi Dornbusch, the Ford Professor Economics and International Management at the Massachusetts Institute of Technology. 'It may be better not to have a common currency at all because mounting popular discontent [especially over the high unemployment of the 18 million Europeans, including the 5 million Germans who have no job] would likely sweep away present policies, including the single currency,'[12] says George Soros. 'I think there's a political elite right across Europe that's for the single currency but is right out of touch with popular feeling. Also, I think it's bad economics,'[13] says Rupert Murdoch. Monetary union is actually a distraction because the Euro will not cure Europe's high levels of unemployment, widening budget deficits, the flight of firms and of capital.[14] For the cure to occur Germany and France must have a free market all across Europe without opt-out provisions by national governments and other tax and regulatory barriers to *laissez faire* trade and investments. Unfortunately, for Europeans hard at work trying to put in place the EMU, ECB and the Euro, these views represent many who judge the whole process leading to a single currency as a Euro-fantasy.

Euro-elite

Thirty-five years ago, the French economist Jacques Rueff made this observation: '*L'Europe se fera par la monnaie ou ne se fera pas*' – 'Europe will be made by its money or not at all'.[15] This is the motto of Europe's political elite. The German chancellor and the French president expect to replace the strong Deutschmark with a slightly weaker Euro. That is their grand EU project; they consider the Euro Europe's manifest destiny. The trade-offs of going forward with the Euro versus postponing or terminating the Euro are both positive and negative for governments, business and workers.[16] Economic pragmatism encourages both going forward and postponing the Euro; however, political pragmatism suggests terminating the Euro, nothing more. The costs of qualifying for Europe's single currency and the costs of postponing the Euro project are high for national governments, central bankers and the bond markets. Moreover, small and medium-size firms, labor unions, workers and small savers in France and Germany might reject the elite's pugnacious optimism about a flexible (and softer) Euro as strong as the current mark. The voters in France might turn out those who would cut down the *acquis*, and the voters in Germany might turn out those who would murder the Deutschmark. In mid-May 1997, the voters in the UK did turn out those who were against additional integration within the EU. British voters rejected American-style (or Thatcherite) free market capitalism in favor of European-style social cohesion in which the 15 nation-states foster market and social responsibilities within the global market context of an expanding European Union.[17] Perhaps, under prime minister Tony Blair, Britain's new interest in modernizing Europe's social democracy within the EU might help the Euro-elite find a *via media* on the road to monetary, banking and currency union – a detour towards keeping both the *acquis* and the strong D-mark within a freer European-wide monetary union. Time will tell whether the Euro-elite, the EU and the whole of Europe can reimage themselves in terms of the 'flexibility' clause – faster and slower groups of countries, a two- or multitiered EU – and still compete against a more productive US and the cost conscious mentality of East Asia.

Is time running out for central bankers? Should they heed the warnings about the impact of the EMU on the rise in unemployment throughout Europe? Can this ascendant governing class preside over the start up of the ECB? Are these finance ministers feeling the pain of longing for a non-national currency they always thought they needed until they finally got it and didn't really want the Euro after all? Will these EU loyalists continue to lead the relentless march of international financial markets through the nation-states of Europe? Can these national politicians think through an acceptable alternative to an EU-wide EMU, perhaps a mini-EMU between France and Germany?

Investment memory

If the Euro is forced on member states that are unable to meet the strict criteria of Maastricht, the EMU might collapse under the weight of high unemployment, slow growth and price inflation throughout Europe. Here is an historical example that sheds light on what might happen to the EMU.

Once the UK was no longer a member of the European rate mechanism (ERM) in 1992, the British government began to pursue a policy of competitive devaluations. This new exchange rate policy becomes the foundation of corporate cost analysis for decisions about foreign direct investments in the UK versus Germany. Successful foreign investors accept a dose of Euro-realism about national sovereignty before they commit to exports, investments, jobs, technology and markets within the 15 nation-states of the EU.

Here's how multinational enterprises get in and stay in a UK that is a part of the EU, but is not a part of the ERM and may not be a part of the EMU. First, invest in the UK. Do not acquire fixed facilities because these factories come with traditional 'old boy' managerial practices and militant labor unions. If at all possible, build 'greenfield' investments. Second, choose sites in the UK, and compare their costs with those in Europe, North America, Latin America and South East Asia. With devaluation of sterling in 1992, employment costs within the UK remain lower than those in Germany and in other countries within the EU whose currencies are tied to the Deutschmark. Finally, in an industry's transition to international expansion, foreign firms need plants within the EU, but they do not need them in countries with high employment costs and those tied to the EMU that is being set up by the political leadership of Germany and France.

Business memory

Since events within the EU shape the responses of member governments to business decisions, business executives must deal with the following changes in their assumptions about the forex value of Europe's currencies.

United Kingdom

- Devaluation of sterling, and its positive impact on exports.
- Export-led growth without wage and price inflation.
- Lack of demand-led growth.
- Reluctance to join the EMU, ECB and the Euro.

Germany

- Revaluation of the D-mark, and its negative impact upon exports.
- Higher employment costs and lost of manufacturing jobs.
- Flight of investment capital to other sites within the EU.
- Push for common currency within the EMU, ECB and the Euro.

France

- Strong franc, and its negative impact upon economic growth.
- High public sector deficits and total government debt.
- Loss of jobs through privatization and social dumping.
- Tie franc to D-mark, and push for the EMU, ECB and the Euro.

Competitive devaluations and a single currency are the stuff of modern international financial management.

Competitive devaluations: very short run

Business managers believe the foreign exchange rate of a national currency reflects market imperfections (such as differences in prices, productivity, interest, unemployment and economic growth). Here are the crucial questions. What is the role of the EU in the global economy? How do its choices and those of its 15 members affect the decisions of other governments, their business corporations, and their financial investors? Can we trace how small changes in Europe's views might have a profound impact upon the US, Japan, eastern Europe, the Middle East and other areas of the world? Who are the key players within the European governments? Which events are crucial to the success of global businesses both within the EU and overseas?

> *'Since the UK was ignominiously expelled from the European exchange rate mechanism in 1992, in a fit of market turmoil and government*

panic, its economy has staged a remarkable recovery. As sterling has slumped by some 20 per cent against the D-Mark over the past three years, its exports have surged by 44 per cent. Meanwhile – and even more remarkably – underlying inflation has remained below 3 per cent for the past 22 months, in a performance more favourable than anything seen for some 30 years. On the face of it, the tale provides a textbook example of competitive devaluation, without the costs of higher inflation.'[18]

Jo Dwek, chairman of the Manchester Confederation of British Industry and the owner–manager of a specialist industrial coatings company, Bodycote, says this about the debate between whether the UK should be in or out of EU's monetary union: 'A competitive [forex] rate is absolutely crucial to UK industry and exporters. If we had a big surge in the rate of sterling I am sure our recovery could be choked off.'[19] During the two years following sterling's absence from the ERM, British exports expanded by 10.9% and imports grew by 6.6% – that is, net exports accounted for almost all the expansion in Britain's gross domestic product (GDP) in 1993 and 1994.[20]

Between 1992 and 1995, Britain enjoyed faster economic growth than the rest of Europe. The UK's current-account deficit narrowed. And inflation stayed relatively low. This competitive devaluation was a short-term success. Why? Britain had plenty of spare capacity. And the British government pursued fiscal tightening by cutting the budget deficit from 8% of GDP in 1993 to 5% in 1995.[21]

Competitive devaluations: medium term

In 1996 the rate of British export growth began to slow, and the rate of export growth from the other weak European currency countries (for example, Italy and Spain) continued to increase faster than for the UK. These results suggest British firms are not increasing their market share overseas. Hence, export-led growth and the free lunch from a competitive devaluation might be over for the UK.

Others are more up beat about the future. Michael Saunders, a UK economist with Salomon Brothers, says 'We have not nearly finished taking the benefits of leaving the European rate mechanism because inflation is still low. It is primarily a loss of growth overseas, rather than a loss of competitiveness which has resulted in the slowdown.'[22] Walter Ellis, formerly chief economic advisor to Britain's trade minister, believes 'UK manufacturers are narrowing the gap in productivity between themselves and German companies with [the latter] now a mere 10 percent more productive than UK manufacturers, compared with 51 percent in 1979'.[23]

British economic history tells a different story. When GDP growth exceeds 3.5%, imports grow faster than exports, and noninflationary domestic demand must fuel GDP expansion. Yet competitive devaluations of sterling against the Deutschmark increase the costs of imports from Germany and other European countries with strong currencies, and might push up the rate of price inflation within the UK.

Thus Britain has two choices for its monetary future. It can manage the forex value of sterling on its own, and pursue more competitive devaluations whenever these become necessary. With luck the UK will not experience price inflation, but this is unusual and it should not be part of future business forecasts. Or Britain can try to manage the forex value of sterling within the emerging EMU, and forego a policy of competitive devaluations. By accepting the conditions for joining Europe's new central bank, the UK gives up national sovereignty over monetary policy. In return, it gets virtually no price inflation, long-term currency stability, membership on the board of the new ECB and, of course, the leadership of Germany's Bundesbank in setting Europe's monetary policy.

Lower labor costs

The UK government's policy of competitive devaluations brings about lower labor costs in Britain *vis à vis* the strong currency countries within the EU. Over time, this way of governing the market encourages business firms from the latter, especially Germany, to invest in the UK. Since 1992, the trend is clear. For example, Germany's chemical manufacturers and car makers are investing more in the UK than firms from the latter country are investing in Germany because costs are higher in Germany.

Hans-Olaf Henkel, head of the Federation of German Industry (or BDI), 'estimates that 300,000 German jobs have migrated abroad [between 1990 and 1995]. Foreign companies hardly ever invest in German industrial jobs anymore. And a growing number of our investments abroad are not based on efforts to expand existing markets or break into new ones but are influenced by the costs in Germany.'[24] More than half of the amount invested abroad by German firms went to five west European EU members. During 1995, Britain, where labor costs are lower than in Germany, received most of the these intra-EU investments.

The following ranks overseas sites for new and additional German foreign direct investment in terms of lower labor costs, a well-trained labor force, a potential for substantial host market sales, an openness to global business opportunities, and, most importantly, a means to improve the profitability of German-owned firms.

Britain

Thyssen, the steel and engineering group, exports jobs to the UK by buying several UK companies that produce castings and pressed metals for the motor industry. Peter Blau from Thyssen says 'It's most definitely a question of exporting jobs. When it comes to making the same products with the same number of people we can make a profit in the UK but are still making losses in Germany.'[25] This is the bottom line for Thyssen.

Asia

German firms are moving production to cheaper sites in Asia because of national requirements for high local content. For example, Siemens strategy is to get into many new high-growth markets and double sales to $14.3 billion by the year 2000. Here are some results:[26]

- *Taiwan*: Siemens broke into the power market by putting together a project finance package of $715 million that out competes similar proposals by GE (US) and ABB.
- *China*: Siemens offers one-stop shopping for subway cars, power plants, and communications networks. Of the US $3.3 billion it will put in Asia,

a third will go to China by the year 2000. According to Klaus Helsen, who heads Siemens in Asia, 'At present there is no money to be made. One must look at the long term.'

Central Europe

German firms are moving production to cheaper sites in Poland, the Czech Republic and Hungary. These three countries plus Slovakia, Slovenia and the Baltic countries (Estonia, Latvia and Lithuania) will grow on average about 4.0% in 1997 compared to 4.4% in 1996 and 3.8% in 1995.[27] The private sector now accounts for more than half of economic activity. Through all the ups and downs, their governments continue with market reforms. Also they have an educated labor force, outwardly oriented economies, and competitive real exchange rates – all characteristics of why the South East Asian countries continue to grow at high rates annually. Nevertheless, foreign direct investment by German and other firms in central and eastern Europe, and the former Soviet Union, an area with 400 million people, equals US $5.7 billion, the amount of foreign investment in Malaysia, a country of only 19 million people.

Although Siemens is looking to improve its profits from its core businesses by investing in Britain, Asia and eastern Europe, the company needs cash to finance its pension obligations back home in Germany. The firm cannot cut costs fast enough nor does its corporate culture permit it to get rid of money-losing businesses quickly. Siemens's profit margins, pretax return on sales, returns on assets employed and free cash available are all lower than its worldwide competitors, GE (US) and ABB. Hence, Siemens will continue to make investments overseas and export jobs to lower cost countries at a pace somewhat slower than its competitors.

The outward flow of investments and jobs from Germany to Britain, other west European countries, eastern Europe and to Asia is steady and growing significantly. The inward flow of investments to Germany from Britain and Europe, the US, and Japan is declining across the board in all manufacturing sectors. These two trends mean job loss for Germany.

International management

As German companies invest outside of Germany, they become less German. This calls for a different business strategy so that their international expansion continues without substantial disruption from competing American, British, Japanese, Swiss, Dutch and French firms. Let's work through the strategy problem as seen by Mercedes–Benz.

In 1995, German labor was between 15 and 17% more expensive compared with other European markets, such as France and the UK, whereas their levels of productivity were approaching those in Germany. Here is how Helmut Werner, the former chairman of Mercedes–Benz assembles the data: 'Rising labor costs, a falling dollar and turbulence with the European currency system have sapped much of the charm and profit out of "Made in Germany". The dramatic appreciation of the mark against other European currencies is just as much a threat to German industry as the falling dollar. The fundamental problem of German exporters is that we are producing in a country with a hard currency and selling in countries with soft currencies.'[28]

Similarly to other German companies, Mercedes must buy fewer German-made components, make fewer products in Germany and expand its manufacturing presence in dynamic growth markets, particularly North America, Latin America and Asia.

For example, in 1995, Mercedes bought 15% of its supplies from outside Germany; by the end of the decade, Mercedes plans to buy 30% or more of its requirements from overseas suppliers. Also in 1995 Mercedes produces only 2% of its cars outside Germany; by the end of the decade, Mercedes will produce 10% of its cars outside Germany. Moreover, virtually all new investment today and tomorrow will go to growth markets overseas.[29]

In 1995 Werner says he needs the US dollar to rise to DM1.60 before Mercedes can make healthy profits on cars exported from Germany to the US.[30] Below that forex rate Mercedes only makes marginal income on its export sales to the US. Even with the US dollar at DM1.62 in early 1997 and with every expectation that the dollar will rise against the Deutschmark, Mercedes is accelerating its investment in the US (in Tuscaloosa, Alabama), and will do the same in South East Asia.

What does all this mean? As Mercedes–Benz moves to become less German, it will be forced to fight a long rearguard action against labor and government in its home country. The former do not want to lose their jobs; and both the local and national governments do not want to lose jobs, taxes and new investments. They will press Mercedes and other German firms to stay put in Germany. This conflict will delay their international expansion at a time when Asian, Latin American and east European markets are open to many more competitors from around the world. Sterling's expulsion from the ERM forced German firms to rethink their business strategy and 'free themselves from the chains German government-business culture had laid on them' for so many years.[31]

The UK and labor costs

Before Britain joined the EU's Social Chapter under prime minister Tony Blair, business executives cut labor costs by creating jobs in those EU countries with loose employment laws. For most of the 1990s, the UK had opted out of the EU's Social Chapter – that is, a common set of rules for working hours and consultations with employees throughout Europe. Since Britain did not accept the minimum EU social legislation to protect workers' rights and benefits, Maytag's Hoover unit believed the low cost option was to transfer its production of cylinder vacuum cleaners from Dijon, France (a loss of 600 jobs) to Glasgow, Scotland (a gain of 400 jobs).[32] With this decision, the firm rationalized its European production in the one site that has excess capacity, and also benefited from sterling's weakness *vis à vis* the French franc and its mirror currency, Germany's Deutschmark.

Most importantly, Hoover cut its direct wage and nonwage labor costs by 25%. Hoover along with Kimberly Clark, Rockwell International's Graphics Systems and others picked free-market Britain against more heavily regulated France. For example, nonwage labor costs were about 50% of wage costs in France, but only 15% in the UK. This difference came about because employers in many continental European countries paid for the health care costs of their employees while their competitors in the UK work under a system where all citizens bear these costs through general taxation.[33] The French called this transfer of jobs to the UK, social dumping, and the EU commission, under its previous French president, Jacques Delors, called it job poaching.

Here is the reality. Even with British membership in the EU Social Chapter, 'A skilled electrical worker in France or Britain costs 40–60% less per hour than his German counterpart, and in Eastern Europe 80–90% less'.[34]

Of course, the shift of jobs from Germany and France to the UK is a does of Euro-realism. 'Ten years ago, the key factor for a manufacturing plant was the size of the national market. Now it's a question of the European market. Previously, companies might make products in three countries to cater for national standards and tastes. Now, the point is that economies of scale can actually be achieved'[35] by producing in one, low-cost EU country, says David Reeves, a location specialist with Ernst & Young. Nevertheless, unless the EU nation-states benchmark themselves against the high growth, low labor cost countries of East Asia, none of them will be as prosperous as they can be. This means Europe will remain in third place behind North America and East Asia.

Productivity and EU cohesion

During the 1990s, the north–south split within the EU resurfaced once again among European-owned multinationals. British, German, French, Dutch and Belgian executives are more doubtful than Italian, Spanish, Portuguese and Greek executives about the possibilities for economic union, and the long-term benefits of Europe's single integrated market. This same division between the potato and spaghetti belts comes out clearly in how they view monetary union, but with a significant difference. Some in the north are more skeptical with over 55% of German and Dutch business executives firmly in the camp that does not see an agreement likely by the year 2000.[36] On the other hand, only about 40% of British, French and Belgian executives agree with their German counterparts. So this major German-led EU initiative remains an elusive goal even for business executives from the European country most committed to the EMU, ECB and the Euro.

Another split among business executives occurs when they speak about Europe's future competitiveness. The UK is seen as the most successful

EU country in cutting costs, shedding staff, and making itself over for the future. Notwithstanding less stringent makeovers in Germany and elsewhere, over 40% of Dutch, Belgian and British executives believe Europe will slip behind East Asia (especially, China and Japan) and the US.[37] Also German, Belgian and Dutch executives see emerging market economies in east Europe as the greater challenge.

American business executives 'acknowledge that, in sectors such as power generation equipment, chemicals and retailing, Europe remains competitive. However, Europe … [lags] badly in many important high technology industries, including computer hardware and software, data networking, and multimedia telephony and entertainment.'[38] These Americans comment that many European car manufacturers, who have been protected from Japanese competition, are well behind Ford and General Motors in Europe in improving productivity, cutting employment costs and shedding managerial staff. Some Americans acknowledge that new investments by Mercedes-Benz and BMW in the US, and additional investments by Volkswagen, Renault and Fiat in Latin America and east Europe are designed simply to escape high domestic costs in Europe. Most American executives, especially in the electronics, computer-based telecommunications and other high-tech industries, know relatively little about European companies because so few of the latter are good enough to set international standards in their sectors and industries.[39] The European rush towards compliance with ISO 9000 standards of quality and performance might change this American perception by the year 2000.

Competitive advantage and exporting

Are European multinational firms falling behind their US and Japanese competitors? 'While high technology products account for nearly a third of US exports and more than a quarter of Japan's, they make up less than fifth of those from west Europe. Europe's weakness in computers and office machinery … shows indirectly by the export/import ratios of science-based industries, with a ratio falling from 0.54 in 1985 to 0.39 in 1990 (compared to 3.96 for Japan and 1.65 for the US).'[40] On the other hand, Siemens, Europe's largest electronics group, is developing powerful memory chips for the telecommunications and automotive markets. Philips, Europe's second largest, is producing liquid crystal displays for portable computers. And Ericsson, after investing 22% of its sales revenue on developing new telecommunications products, controls 60% of the world market for digital mobile phone systems.[41] Also Europe seems to be holding its own in drugs, medicines, and biotechnology – all areas of traditional strength.

Is third-place Europe's future? Probably yes, but Europe will not be as far behind tomorrow as it is today. The northern countries, especially Germany, The Netherlands and Sweden, are intent on running a good race with the Japanese and Americans; the UK, France, Belgian, Denmark, Finland and Austria will not be too far behind their northern European competitors. Nevertheless, the southern countries (Portugal, Italy, Greece and Spain) will drag down EU averages for high technology exports and R&D spending on civil science. These may be overcome somewhat by additional high technology strength from eastern Germany (for example, Siemens investments in Dresden), western Poland (for example, ABB investments in power generating equipment), and the Czech Republic (that is, Siemens, VW and other German firms).

Competitive advantage and privatization

What happens and how fast it happens depends upon how quickly governments sell off state properties, and whether privatizations promote efficiency and competition. As always in Europe, national goals vary and methods differ from country to country. If we split up state-owned monopolies before sale, this tends to reduce their value and the price private-sector buyers are willing to pay for state enterprises. If we privatize the whole monopoly quickly through a mutual fund sale and vouchers distributed to the public, these efforts tend to reduce the political tension over the sale of key state-owned firms. Of course, financial markets judge success in privatizations in terms of productivity, efficiency, product development and competition.

Here are some examples of European privatizations:[42]

- UK (BT and BP), Germany (Deutsche Telekom and Lufthansa) and Sweden (Televerket) are privatized to induce competitiveness in crucial industries, such as telecommunications – a politically deft maneuver if it does indeed deliver better services at lower prices to local customers without reserving 'blocking shares' to local governments.
- France (France Telecom, Rhône-Poulenc, Renault, Thomson-CSF and Elf Aquitaine), Italy (STET, ENEL, ENI, Agip and Finmecanica) and Spain (Telefonica, Repsol and Endesa) are privatized at bargain-basement prices to cut the public sector's deficit – a politically uncertain way to do business in the face of nationalist out cries from labor and the public.
- As the east European countries shift from the command to the market economy, the Czech Republic already sheds its heavy industry and Poland privatizes over 40% of the national economy. And Hungary leads the way in maximizing foreign ownership throughout all of its national economy – a politically uncertain way to do business in the face of resentment from local voters.

If these privatizations work, then many of Europe's industrial giants will stop accepting state aid in return for giving their employees civil-service-style job protection. Their management will be free to take market risks and convince the world's capital markets to finance these choices at home, elsewhere in west Europe or overseas. However, labor markets at home will remain as rigid as ever, scuttling most opportunities for real reform.

Are Europe's industrial sectors sufficiently competitive to outperform the same manufacturing industries from Japan and the US? 'Output of US manufacturers between 1980 and 1992 rose twice as much as that of the EU, while Japan's rose far more. Output per person in EU manufacturing also rose less than in the US and Japan. One explanation for the poor performance of manufacturing exports must be the EU's relative emphasis on internal rather than external liberalisation. The EU has, in effect, been pursuing an import-substitution strategy, one favouring internal sales over exports. A second explanation would be a mixture of inadequate innovation with an unfavourable industrial structure.'[43] Since Europe specializes in exporting goods with less technological sophistication than those from the US and Japan, Europe is forced to compete on the basis of price, not unlike the problem facing exporters of manufactured goods from Central and east Europe, Latin America and South East Asia.

The saga of *Mittelstand*

When sterling plunged against the D-mark between 1993 and 1995, German firms were slow to take investment risks abroad. There are good reasons for this hesitation.

Many of Germany's big firms are unwilling to face strong government and labor opposition to moving jobs abroad, even when they go to other EU countries or nearby east European countries. Moreover, who could tell how long sterling would remain cheap *vis à vis* the Deutschmark. That it went on four years and caused BMW, Mercedes–Benz, Siemens and others to build new factories in the UK and the US gives rise to other concerns, especially the weakening of ties to the home country. The strong Deutschmark and the competitive devaluations of other governments give these big firms the opportunity to flee high German employment costs. Today, Germany must learn to deal with German-owned multinationals that are more immune to political pressures from the German government.

Most of Germany's medium-size firms (*Mittelstand*) already are world class because they are export-oriented and technologically smart. They see no need to invest overseas. If they show an interest in doing a joint venture with a foreign firm, these mid-size German firms do not have the equity to invest overseas. Unfortunately, venture capital to grow new high-tech initiatives or new overseas markets is just not readily available in German financial markets, and if it were, it would be taxed at rates far higher than those found in the US and the UK.[44] So these *Mittelstand* firms export goods. However, they stay home with their master craftsman talent, family management and hidden capital. As German employment costs rise, these medium-size firms see a future in which their products are uncompetitive because of the strong Deutschmark. These firms are the victims of the competitive devaluations implemented by other European governments.

Yet a majority of small and medium-size German businesses do not want to give up their strong Deutschmark for a European currency that might or might not be as strong as Germany's national currency. They fear inflation, more public-sector debt and a central bank unwilling to say no to politicians. Are their fears well grounded?

Many weak currency countries will be founding members of the EMU. So German firms must worry that their public debt strategies will sink the Deutschmark and the Euro. If this becomes the major view of Euro-realism, then the EMU becomes unfeasible. Of course, the loss of the EMU, ECB and the Euro initiative is politically unacceptable to Europe's leaders. Therefore, German business executives have a right to worry about the future of their Deutschmark as it gets subsumed with the French franc (and maybe the pound sterling) within the Euro. This too is Euro-realism.

Lessons about currency fluctuations

Until the EMU, ECB and Euro start up, in 1999 at the earliest, business executives will have to watch their forex exposure. Germany sends 55% of its merchandise exports to other EU countries, and 45% to non-EU countries, such as the US, Japan, South East Asia, Central and eastern Europe, and the Middle East. France ships 53% of its merchandise exports to other EU countries, and 43% to non-EU countries, such as the US, Japan, North Africa and the Middle East, South East Asia and sub-Saharan Africa. Over 40% of the exports of Germany and France are exposed to large currency fluctuations. The EMU will not give these two countries currency stability (with their national currencies or with the Euro). If intra-European currency conversion is eliminated between both national currencies, this might add 0.4% to the GDP of the EU, not enough to warrant the costs of setting up the EMU in the first place.

Sovereign risk

Under the EMU France retains its sovereign risk to restructure its public sector debt and pay it back in Euros not French francs. Also under the EMU, France retains the virtual independence of its national fiscal policy. Given Germany's backing of the Euros, France can refinance its sovereign debt at lower rates. Therefore, France and other weaker currency countries get a windfall gain by joining the EMU, and using the latter's financial conventions to carry on much as before with its labor unions, government monopolies and a subsidized private sector. Nation-states remain sovereign under the EMU. They must agree to fiscal policy integration and a uniform tax policy across the EU. Otherwise, the EMU provides no bankable guarantees against erratic economic behavior from Paris, London or Rome. That is the great fear of German business executives.

Nevertheless, GE Capital, the financial services subsidiary of General Electric, the US industrial giant, is stitching together one of Europe's broadest financial groups. In Europe, GE Capital has made some sizeable acquisitions, including credit card businesses of UK retailers Burton and House of Fraser; the vehicle fleet services business of Avis Europe; the vehicle finance, computer leasing and office equipment leasing companies of Skandinaviska Enskilda Banken; a French vehicle fleet management company; and an Austrian consumer finance business.[45] In 1995 alone, GE capital spent $3 billion to buy Germany's Frankona Re and Aachen Re, France's Credit de l'Est, Britain's Pallas Group, and Poland's D. T. Bank.[46] All of these acquisitions make GE Capital a low-cost provider of financial services within Europe.

According to GE's Italian-born vice chairman Paolo Frecso, 'We forecast that European sales would double by the year 2000, from $10 billion to $20 billion. Now some think the $20 billion figure is too low.'[47] Investing in Europe's slow growth market is a mammoth, contrarian bet on the world's most mature market which gave GE in 1995 a substantial increase in operating profit while its Asian operations showed a small decline in operating profit.[48] Europe is making money for GE. This too is Euro-realism, of course, with an American challenge.

Identity citizenship

'I come from Lorraine, my culture is German, my nationality is French, I think in our provincial dialect,'[49] and I work for an American or a Japanese firm. Today, Europeans (as well as some global-trotting Americans, many English and Scottish kit and kin in the British Commonwealth, most overseas Chinese in the Asia–Pacific region, and all Mexicans from Durango and Michocan states in Mexico who live in Chicago) have multiple, combined identities – passports in one hand, and networks of family relationships and personal connections in the other hand. None want to be detached from their past cultural practices, but none want to give up the concrete advantages of classifying oneself within a larger group in order to enjoy the extra access of European citizenship, jobs in the US and investment opportunities in South East Asia. Europe is living through a cultural revolution in which many national citizens are orphaned because their communities of like-minded nations are hard to find – sometimes at home, elsewhere in Europe and, of course, in East Asia. The English might choose to be English, but they may have to settle for being Anglo-American in Bangkok, Thailand. Or the French might choose to be French, but they may have to settle for being Western in increasingly English-speaking Vietnam, and Zaire (Congo) and central Africa. The Dutch, Japanese and the overseas Chinese hold strong beliefs about their group identify; however, they have no choice but to speak English and enjoy American cultural habits whenever they are far away from home.

In the U-World of loyalist nation-states everyone is both insider and outsider. Their skin is Welsh, Bavarian, Moravian, Tartar, Mung, Mohawk, Maya, Zulu and Afrikaner. Their shirt is British, German, Czech, Russian, Laotian–Vietnamese, American–Canadian, Mexican–Guatemalan and South African. And their tailored coat is EU, and the ill-fitting outer garments are NAFTA, Mercosur, Asean and APEC. Identity citizenship is the social construction of like-minded people who come from diverse ethnic and national backgrounds. In some places, they are Bavarian–German–European, or Tyrolean–Italian, Slovenian and Austrian–European – all acceptable identities that are interchangeable as time, place and custom permit. In other places, they are Latin Catholics, Greek Orthodox, Turkish Muslims–citizens of Turkey–European, too, but simply Italians, Greeks and Turks in western Europe and the US. The place, time and context play an important role in whether we use nationality, family or identity to signify our place in the global, largely American market economy.

Japanese business executives generally reject ill-fitting outer garments and cling to their own ways of doing business in Japan, South East Asia, the US and Europe. American business executives are uncomfortable in ill-fitting outer garments when they are overseas, and prefer to do their work and run their businesses as they did back home, but they will make some adjustments to Canadian and British attitudes about managerial practices. And so will the British when they are in the US. Continental Europeans, especially the Germans, Swedes and Italians, make themselves as comfortable as possible in eastern Europe, Brazil and Argentina. The overseas Chinese are at home in Australian, Malay, Thai, Shanghai, Canadian and American garments so long as there are other Chinese to work with locally. A win-win world is built on skin, shirts and outer garments all fitting nicely on the backs of crucial business executives.

Win–win–lose world

The U-World is at the beginning of a win–win–lose EU era for foreign investors in the UK. Both American and Japanese firms foresee business opportunities in the UK because this national economy offers unrestricted access to the EU. Foreign direct investments in the UK are not simply about domestic manufacturing or exporting to Europe, the Middle East and Africa. Today's EU-based companies have to be cost-competitive on a global basis with integrated systems of sourcing from the home country, within the UK and across Europe. Their local employment and infrastructure costs must be competitive with comparative expenses elsewhere in Europe, in North and South America and in East Asia. Hence, the recent addition of German firms to those foreign firms seeking out business opportunities in the UK.

Since the UK, especially England and Scotland, has been a favorite of American firms for almost 50 years whereas Japanese firms only started coming to England within the last ten years, a great deal of evidence exists about their relative success in managing overseas subsidiaries. The American firms acquired locally owned manufacturing facilities or built new factories, and became forever identified as British without a European passport. On the other hand, the Japanese firms preferred to put down new 'greenfield' investments; and they stayed Japanese in corporate culture, work habits and social connections with local people. Both Americans and Japanese viewed the use of English among the local population a definite plus. The Americans were willing to over look traditional 'old boy' management practices and militant unions. The Japanese preferred to alter the former, and they benefited from the changes in labor practices brought on by the conservative governments of prime minister Margaret Thatcher in the 1980s.

As a consequence, many Japanese firms thrived in the UK. Here is a short list of what they do differently to their American competitors.[50] First, they convey to their employees long-term job security. Second, they use problem solving by teams rather than managerial bureaucracy to create a flexible work setting. Finally, they want to be market leaders in the UK and Europe. Nissan is an example of a successful Japanese transplant to the UK.

Unsuccessful Japanese- and American-owned subsidiaries in the UK (such as Toyota and GM/Vauxhall) were unable to get production up to levels where manufacturing had economies of scale for the European market. Some American firms followed traditional British labor practices by laying off workers (or redundancies) when things got tough. These Japanese and American firms were unable to build market share because they did not have clear market segmentation and positioning strategies. Nor did they stress quality, service and innovation.

In the mid-1990s, a new set of competitors has entered the UK market. These are German-owned subsidiaries who bring with them talents in building high-valued added products and stressing quality and innovation. They come to the UK because employment costs are competitive in terms of the forex value of sterling.

Euro-realism

Here we have Euro-realism in a nutshell. Money (that is, the Euro, the forex value of national currencies and the labor costs of manufacturing) always wins within the EU. Adopting a single European currency implies running national economies at high levels of unemployment to meet the economic criteria set by Germany.[51] The situation for France, Italy and Spain is unstable. Because different national cost structures and dissimilar international trade patterns arise from the very heterogeneity of the 15 member states of the EU.[52] Therefore, it is unrealistic that all these countries can be made to conform to one standard set by Germany under the veil of the EMU.

Is there a way out? Perhaps in 1999 only Germany and France should form a mini-EMU with or without the strict criteria set down by the Maastricht Treaty.[53] Perhaps the best idea is to set the deficit criteria aside by talking them to death. The independent Bundesbank (or together with an emerging ECB in which all EU central bankers might be *ex-officio* members) could manage how and for how long the German Deutschmark does a leveraged buy out of the French franc. Also the Euro itself might not be used until all EU countries (especially, Italy, Spain and Britain) are finally in the EMU. Moreover, the French and German governments could impose budgetary and fiscal measures to bring their national cost structures in line with the expectations of international financial markets as they price up or down the future forex value of the Euro. A two-step approach to putting in place the EMU is right for today's Europe.

Money always wins. But its victory must be tempered by *la douce France*: the deep, old France that begins outside Paris; the France of small villages and gentlemen farmers; their consuming hatred of 'money power' – whose real home is not in France but Somewhere Out There; the belief in government's ability to counter money power with national power.[54] Money's victory must be moderated by the realities of Europe's different national economies, business performances, labor cost structures and international trade patterns. Otherwise, national governments (with or without the approval of Brussels) will impose high transactions costs on capital flows, short-term 'hot' funds and longer-term direct investments. Today, German boldness together with *Le Défi Français* means monetary union is first and foremost a political matter for the permanent Franco-German alliance. Understanding this crucial idea is the first step towards accepting Euro-realism.

Costs and benefits

How should Europe weigh the costs of giving birth to the European Monetary Union against the benefits of less currency conflicts in the future?

This is a case of decision-making under uncertainty. If only France and Germany join the EMU at its inception, both countries might suffer higher unemployment, but other EU nation-states may gain jobs as their national currencies decline in value *vis à vis* the new Euro. It is possible the forex value of Euro would remain stable against the US dollar and the Japanese yen. On the other hand, if all the European countries involved joined the EMU at its inception, the possibility of higher unemployment, slower growth and greater price instability loom ahead for the ECB; these results could cause the forex value of the Euro to fall against the US dollar and the Japanese yen. How much a loss is unknown.

Decision theory cannot create certainty where none exists, but it offers a way to organize one's thoughts and gather the information needed to make better decisions. Consider monetary union. The best policy is for a few member states to join the EMU, sit on the board of the ECB and monetize the Euro.

Remember Europe has choices. One option is to deepen the EU through a two-step approach to the EMU. France and Germany enter first and seal their post-World War II partnership with monetary union as a grand act of political will. Other northern countries and some or all southern countries join the discussions as *ex-officio* members of the ECB, and when they can provide national budgetary support for their levels of unemployment they could join the EMU, too. When all those who are going to join the EMU actually do, then the Euro could be declared the cross-national currency for the EU.

A second option is to go for broke and do everything to set up the EMU, ECB and the Euro by the Maastricht target date of 1999. Government leaders think they can pull this off. However, higher unemployment and the loss of 'acquis' means Europeans grow in their anger for the EMU, and in their ambivalence towards the EU itself. It would be a shame if the EU stumbles badly because of haste to include all EU nation-states in the EMU.

Today, the EU is pursuing option two without regard to the true downside costs of the EMU. Will Europe be able to continue with this unhappy state of affairs? Probably yes. Because so many national leaders have staked their political reputations on initiating the EMU, the ECB and the Euro on schedule in 1999. However, higher levels of unemployment, slower growth and greater price instability might give them some pause in their pell-mell rush towards full implementation of the EMU. Late in 1997 or early in 1998 might be the time of surprise in the debate over the EMU.

Decision theory offers us a way to reflect on our choices and make the best decision we can with the information we have in front of us. Decision theory offers Europe a way to organize itself to beat the fate of Salmon Day.

Salmon Day

Salmon Day is the experience of spending an entire day swimming upstream only to get screwed in the end. Many American, some Japanese and a few German firms invested in Britain and gained market share in Europe. They took advantage of lower UK labor costs that came about because Britain was outside the ERM, Social Chapter and the emerging Euro. Of course, those foreign firms that failed did so because they were poorly managed. They are victims of Salmon Day.

Here are some summary thoughts. First, the transmitters of Europe's common knowledge about the big idea called the European Union offer the following: participation in the Franco-German approach to the EMU, *ex-officio* or voting membership in the management of the ECB and a cross-border, non-national common currency called the Euro; relief from competitive devaluations as a source of jobs, investments and exports; and help in price stability, economic growth and more employment.

Second, the editors of Europe's common knowledge are in the process of reviewing the start-up costs of the EMU, and speculating on how a two-step approach might avoid the pain of political failure and minimize the true economic costs to the nation-states. Some of their vivid memories includes the expulsion of the pound sterling from the ERM, and how the UK used a competitive devaluation of the sterling against the Deutschmark to encourage German investors to create jobs in Great Britain.

Third, managers and others in the professional class must decide whether the single currency is more important than an expansion of exports and imports. Money is homogeneous and fungible, and transactions occur instantaneously across countries, common markets and the world. On the other hand, trade in goods and services is heterogeneous, and transactions occur discontinuously, mostly between nearest neighbors (such as France and German, or Canada and the US) and among best trading partners (such as the US and Europe, and the US and Japan). To decide for the single currency means Europe runs the chance of disrupting trade patterns within the EU for one or more years. However, to decide against the single currency means Europe runs the great risk of stopping the EU in its tracks. Already, polls exist about whether the EMU will happen and whether it will be a success or failure. These questions will increase in number as Europe moves closer to the final date for initiating the EMU, the ECB and the Euro. Here then is Europe's danger as it swims upstream with the probability of getting screwed in the end.

Wisdom: decision theory lays out possible choices. Pay attention: one choice leads to Salmon Day. The other slows the process of single currency integration and might avoid Salmon Day. Open the doors: Europe makes a choice to go forward with a two-step approach towards initiating the EMU.

What do I get?

Here is our to do list:

- Buy into Europe's big assignment for the EMU.
- Accept *ex-officio* and voting status on the ECB.
- Build on the memories of past competitive devaluations.
- Reimage ideas about a one-step versus a two-step approach to the EMU.

- Recognize low labor costs attract foreign investors within the EU.
- Create more trade in goods and services with or without the Euro.
- Use decision theory to weigh costs and benefits of options.

Remember these words: Salmon Day for the EMU can be avoided, but France and Germany will have to rethink their approach towards initiating the EMU to be sure Salmon Day is avoided for the EU.

Chapter 6

The East

'If the European Community has succeeded in surviving and, even, in making some progress, this has always been at the price of maintaining a persistent ambiguity as to its ultimate destination.'[1]

Valléry Giscard D'Estaing, former French President.

Europe is 'outgrowing the concept of the original founders of the Union'.[2]

John Major, former Prime Minister of the UK.

Inventing Europe

The 1990s are the decade of the Europhobe.

The Europhobe is frightened and angry. Fear transcends traditional social, cultural, economic and political boundaries. Fear generates unity: French farmers join forces with German food merchants; ex-hippies take up arms alongside Protestant clergy; the European Commission stands shoulder-to-shoulder with right-wing nationalists. A Europe bitterly divided on an array of EU issues from the EMU to the Social Chapter can unite on this: new members from the East – all of them endanger high prices for EU labor, food and manufactured goods.

The Europhobe defines the East broadly. The East is every city, town and village east of Berlin. For sure the East starts where the Latin West ends somewhere in Poland-A (the pre-World War II German lands), the Czech Republic, Hungary and Slovenia (the former Austro-Hungarian

Empire). Of course, Europe is not the Orthodox communities in Russia, Ukraine, Romania, Bulgaria, Serbia, Macedonia, Greece, Georgia and Armenia. Nor is Europe the mixed Orthodox–Moslem communities of Bosnia and the Balkans, and the Moslem communities of Turkey, Azerbaijan and the Caucasus.

Yet for all the clucking by west European loyalists, rooted cosmopolitans, East and West, both want and embrace the new East. Poland, the Czech Republic, Hungary and Slovenia – countries still within the Latin (Catholic and Protestant) West – are among the first prospects to join the EU. The broadening view of the EU has swiftly changed from a futile political movement little loved outside of Great Britain to a mainstream economic mandate of the Franco-German consortium demanded by some of the world's largest corporations.

Notice the apparent contradiction.

Europhobia is exactly what it sounds like: not concern about real problems but an anxiety disorder, an increasingly irrational spiral of unwarranted fears about the unknown lands, cultures and peoples in the East.

This cultural conflict, the endless negative feelings about whether the lands to the east and north of the Bosporus belong to the self-centered and world-dominating outlook developed by the lands to the west of the Bosporus, is pointless. East and West, perpetually at war, have a kind of reflexive arrogance in common: each group sees itself as superior to the other. There is, of course, no good reason for us to have to choose between the Latin West and the Orthodox–Moslem East. Both are valid and useful in their way, and neither is going to go away. The emergence of the East freed of communism and broader definitions of Europe does not mean that the West will or should decline. But we are groping for some sort of coherent response to all the knocking on doors we are, sometimes literally, hearing.

The European world of the 1990s bears little relationship to that of just several years ago, a time that feels as remote as the Roman Empire. Remember the DDR in whose east German lands begins the vast indeterminacy of Eurasia. Although raised up as a communist success story, the DDR was a Potemkin village from which citizens fled the crushing failure on foot, by train and in their Trabi cars. Today, these new Germans cannot find work that pays them what their fellow citizens get paid in west Germany. During the last 500 years, the geographical and historical concepts of Europe have moved West, and the Atlantic peninsula – that is, the far-western Latin West with its secularism, materialism and consumerism – is what we mean by Europe.

The notion that we in the West have two distinct cultural choices has become widely accepted. If we are broadminded and cosmopolitan, we may think the Orthodox community of the East, with its vast distances and brutal petty officials, is European or not, as we choose. The point is that we have to choose and do not quite know how to choose, and the peoples in East have the corresponding problem from their own perspectives.

Sensible choices

From any distance, the construct of such narrow choices seems pointless. Why should East and West be defined by out-of-touch loyalists, officialdom and clergy? Why can't we make sensible choices that draw from different elements of Latin and Orthodox culture, choices that further the integration of Europe, East and West, that fit into our lives?

We can, of course. Sanity begins by relying on common sense. Enlightened people will educate themselves about the European Common Market, Economic Community and Union – to list the names it has successively called itself. And how it has successively resolved coal and steel overcapacity, agricultural supports, free trade in goods and services, mobility of labor and capital, and, perhaps, the single currency.

First, sensible Europeans will figure out it was and has remained a Franco-German consortium, asking democracies to give up their sovereignty, inducing them to take the first step and then revealing to them that it was irrevocable, so that no way remained but forward. The separate histories of the British, French, Germans, Spanish and Italians do not work anymore; soon neither will those of Poles, Czechs and Hungarians; and in the future the histories of the Russians, Ukrainians, Rumanians and others will merge into something called the history of Europe.[3] Since the separate cultures cannot guide us, we have to make our own way towards a blending of national cultures into a globally, largely American (and partially European) consumer culture.

Second, sensible Europeans will open the door wider towards those areas in Europe in which the Catholic–Protestant–Latin West shades into the Orthodox–Moslem–Communist Eurasia. They will work with ancient original 'Europa' (Thrace, Macedonia, Illyria, the more modern Bulgaria, Albania, Serbia, even Greece) now known as the Balkans (from the Turkish word for mountains) to secure the marchlands to the east for Europe, the EU and European business.[4] If the eastern Europeans take the first step into the EU, they will be bound irrevocably to a postmodern Europe in which the global market demands nation-states give up their capacity for self-government over trade, investment and monetary issues.

Through all the changes in the East and the West, sensible persons will recognize that 'the huge bulk of population in west Europe, modernized, bureaucratized, and bourgeoisified, lives in conditions indistinguishable from those of the United States and shares similar tastes and interests. The classless and international modernity … has engulfed Western Christendom and is being hungrily embraced by our Eastern cousins.

'[However,] the surface features of their modernized lives conceal profound cultural differences'[5] between British and French, French and German, German and Pole, Pole and Russian, Russian and Turk, Turk and Greek, Greek and Serb, Serb and Slovene, Slovene and Italian, Italian and Spanish, Spanish and Irish, Irish and British, and European and American. Taken together, the EU and its expansion to the East offer rooted cosmopolitans and loyalists alike a concern about real problems, and a rational way to spread secularism, materialism and consumerism eastward. All of us are groping for some sort of coherent response to the global market without becoming Euro- or EU-phobes.

According to John Major, the former prime minister of the UK, Europe is simply 'outgrowing the concept of the original founders of the Union'.[6] The EU is trying to move away the sovereignty of nation-states towards a more integrated union in which sovereignty is vested in the EU itself. Although the Franco-German consortium might be in favor of this drift, Britain and others are strongly opposed to the demise of nation-state sovereignty within Europe. Their strength breeds Europhobia.

Europe in the making of the world

Europeans have always viewed Europe in the East as a fight between nomadic raiders from the Eurasian steppes and settled civilizations along the borderlands between East and West. Here are some examples of victors and vanquished:

- The Lombards and Visigoths over run France, Spain and northern Italian, and sack Rome. The imperial center moves East to Constantinople (modern Istanbul). The Slavs. march south to the Adriatic and east to Russia. Then the Arabs ride north, conquer Persia's Baghdad, and the Christian lands of North Africa, Sicily and Spain.
- The Mongols (or Tartars) sack Kiev, take Moscow, and over run villages from Kraków in the north to Buda and Pest in the south. Their attacks push others from Central Asian (modern-day Bulgarians and Hungarians) ahead of them who stay in the West.
- The Normans take back Sicily, anchor it to the Latin West. and lose it to Bourbon France and Spain. The Turks, who were converted to Islam by the Arabs, wipe out Constantinople's Christian East Roman-Byzantine Empire, and convert Slavs, Greeks, Armenians and many other people to Islam.
- Combined attacks by Swedes, Czarist Russians, Prussians and Hapsburg Austrians erase Warsaw's Poland-Ukraine-Lithuania Kingdom. The western Allies defeat the Central Powers twice and terminate Vienna's Austria-Hungarian Empire, Prussia's Imperial Germany, the Sultan's Ottoman Empire, and Hitler's Third Reich. Czarist Russia, Poland, East Germany, Czechoslovakia, Hungary, Romania and Bulgaria all fall into the grip of Moscow's Communist army.

Between 1989 and 1991, the 70-year communist empire fell. Lithuania's nationalists, Poland's Solidarity Union, Czechoslovakia's 'Prague Spring' reformers, East German Christian socialists and Yeltsin's democratic Russia topple the rule of the Communist Party, one of the shortest empires in the history of the east–west borderlands.

In its place comes a Franco-German consortium, the European Union, inspired by America's interest in free trade, Europe's desire for peace not war, and the pragmatic pursuit of a better life by all the people in the Eurasian borderlands. Where does 'The East' begin? Where society does not benefit from the Renaissance, Reformation, industrialization, democracy and the international economy.

In Germany's rush to drive the Russians from the borderlands, most Europeans lose sight of the East. They transmute eastern Europe into a stereotyped view of Commissars bungling plans and production schedules, stealing control over state assets and paying lip service to social cohesion among atheists, Orthodox and Moslems. Both Europeans and Americans accept the new monolithic culture from Moscow and refuse to challenge Communist statistics about success and failure. After 1989, both the French and the Americans have a failure of insight about the momentous nature of the changes occurring in the East.

The rise and fall of the former Soviet Union means the opening up of the brilliant cultures of central and eastern Europe to the newer and more dynamic global, largely American (and partially European) life style. Will the east–west borderlands be able to mount a consumer revolution comparable to that which occurred in America, Europe, East Asia and Latin America?

Our big assignment in this chapter and throughout the book is trend spotting. What is faith? What is popcorn? Whither Europe's role in the East.

Corrosive effect of orthodoxies

Today's Europe is a foreign country to Jean Monnet, the father of the European Coal and Steel Community. It is passively modern, sometimes moving and changing, constantly fretting over America's role, not yet able to grasp the nettle of Europe's next historical drama, the linking of the Orthodox East with the Latin West. The Europe of his mind's eye – fully sovereign nation-states linked together to pursue free trade – is gone forever.

According to German journalist Jose Joffe, writing in 1990, 'Western Europe has chosen a path that knows no precedent. It is not political will that fuels the engine but economic necessity. Economic forces – the need for economies of scale or for international competitiveness – are supposed to lead the way. [Such a strategy could succeed only if the member states were willing] to merge their sovereignties into something that is more powerful than each and all.'[7]

And according to German novelist Peter Schneider, writing in 1997, 'We have all been supporters of Europe for a long time, but we are finally realizing that the Union is becoming a reality, and we have no idea what this means culturally, for everyday life'.[8]

Today, the EU bureaucrats in Brussels act as if all history has occurred just to let them make their appearance in the world.[9] They measure every scheme for integration by their own framework, and apply a single measure, dividing central and east European countries into 'suitable' and 'not suitable.' In this fashion, they may let in the last few nation-states of the Catholic–Protestant Latin West and keep out all the countries of the Orthodox and Moslem East.

Also 'the drive to create an economic superstate in Western Europe, specifically the monetary union beginning 1999, could ... leave Central Europe out in the cold. If a monetary union is launched on schedule, it will touch off a chain reaction that will halt integration,' says Jan Krzysztof Bielecki, the former Prime Minister of Poland.[10] The poorer countries of the East might get trapped in a chronic recession because their poorer peasants from east Poland, Belarus, Russia and Ukraine cannot migrate freely to the richer West.

Moreover, Germany, Britain, Belgium, Luxembourg and the Netherlands do not want to come up with the additional $50 billion (in US dollars) for the price supports necessary to keep farmers afloat in Central Europe. They already pay $60 billion in price supports to farmers throughout the EU and in transfer payments to the less affluent member states – such as Portugal, Italy, Greece and Spain – and to east Germany.

As Margaret Thatcher, the former British Prime Minister, said at the first congress of the New Atlantic Initiative in Prague in 1996, European monetary union is 'in truth a nightmare.' But as 1999 approaches member-states of the EU are finding it ever harder to ignore the risks[11] – lower economic growth, higher unemployment and price inflation in the West and chronic recession, vastly higher unemployment and a pull back from free markets in the former communist East.

All this is unfortunate. What would Jean Monnet say? 'My intellectual inheritance is being confiscated by the corrosive effect of today's EU orthodoxy.' What will happen? The EU will declare Polish hams, Czech tomatoes, Hungarian wheat and Slovenian wine – and all their workers – a clear and present danger to the EMU, ECB and the Euro.[12] That is putting the cart before the horse as central and east European countries knock on the free trade door of the EU.

Let us start with East Germany and see how its reintegration into Europe worked out. Then we will outline the particulars of integrating 'suitable' countries of the Latin West. Finally, we will check off why the countries of the Orthodox and Moslem East are not 'suitable' for membership in the EU. In each case, we will ask this question: Does EU orthodoxy (such as free trade in goods and services, free mobility of labor and capital and the single currency) improve or diminish the chances for admitting new members from central and eastern Europe? What follows is Europe's road map to a future waiting to be written by the Franco-German consortium that leads all new EU initiatives.

Wende of *1989–90*

'What did they imagine?' West Germans always raise this question about their conationals from East Germany (the former DDR). After all, the east was poorer than even the poorest countries in the EU.[13]

Seven years have gone by since the communist east was taken over by the market economy of West Germany. The *Wende* (turnround) happens speedily as all the inefficient state enterprises are sold off by the Treuhand, a special government agency for the former DDR. Also the East German mark is revalued on a 1:1 basis to the Deutschmark. This gives East Germans the purchasing power required to buy all the Western food, clothes and cars they desire but could not afford in the past. Moreover, their wages and welfare benefits are raised to Western levels, too; unfortunately, their productivity levels do not go up as fast so their manufactured goods are not competitive in German, European and world markets. Finally, the telecommunications and transport infrastructure are rebuilt and modernized, and the housing stock is privatized.

'Unemployment at about 15 percent is 50 percent above the national average. Manufacturing accounts for only about 18 percent of gross domestic product (GDP) ... well below the 25 percent that would be normal in a modern industrial region.'[14] German transfers to East Germany make that country the European champion in granting subsidies to one of the poorest regions in the EU.

The result: the East Germans still have high levels of unemployment, vast economic uncertainty and no self-sustaining upturn; and the West Germans still must subsidize jobs, wages and welfare benefits in the former DDR.

Zweckpessimismus *of 1997*

'Is the economic catching up process coming to a halt?' Yes says German economics minister, Günter Rexrodt, in his report to the president of the European Commission, Jacques Santer.[15]

This deliberate gloom (*Zweckpessimismus*) means the new *Länder* need more subsidies from Germany and the EU. The capital stock of their state enterprises is worn out and need to be junked by the Treuhand. The new exchange rate makes exports from the former DDR to the former communist countries of eastern Europe too expensive. The new parity wages and welfare benefits with West Germany price the labor from East Germany out of the EU market. Even worse the construction boom in infrastructure, and commercial and residential building is coming to an end without any alternative jobs opening up in East Germany.

Currently, West Germany transfers about 5% of its GDP to the east. Also German individuals and corporations pay a 7.5% 'solidarity' surcharge on their income to fund the rebuilding of East Germany. An end to these subsidies and other welfare benefits means a mass transfer of population from East to West Germany.

The result: the East German growth rate is falling behind the most dynamic economies of Central Europe; exports from East Germany to its east European neighbors keeps declining while exports from West Germany grow by 20% annually.

Vergangenheitsbewaltgung

'Should we continue concentrating subsidies on the industrial core – autos and chemicals?' Yes says the premier of Saxony, Kurt Biedenkopf even when the EU wants to lower the subsidies offered by this *Länder* to VW.[16]

Saxony, Thuringia and part of Saxony-Anhalt have been Germany's industrial heartland under the Kaiser, Weimar, Third Reich, DDR and East Germany. Brown coal, other minerals, chemicals, cars and good communications industrialized Dresden and Leipzig. The Treuhand cut their work force by 20–25%, sold off 15,000 businesses and privatized another 5000.

VW and Opel built highly productive auto plants at Chemnitz and Eisenach, respectively. But East Germany lacks the *Mittelstand,* the smaller and mid-size firms that can supply these new plants and export out of the region to West Germany and to the rest of the EU. There are no specialized and niche manufacturers in East Germany's south.

The new *Länder* from East Germany have reduced Germany's GDP per head. Once Germany only trailed the Swiss and the Japanese. Now it trails the Austrians and Belgians as well.

The result: East Germany needs more subsidies rather than less, but it might get less rather than more from West Germany and the EU; because, Germany needs to cut its budget deficit to meet the strict Maastricht criteria for joining the EMU. *Vergangenheitsbewaltigung* or the need to come to terms with the past, adapting to the present and getting ready for the future still overwhelms both East and West Germany.

Is the country as a whole losing dynamism as its gives out subsidies and falls prey to narrow interest groups, such as those from the south of East Germany? Or should West Germany give out more subsidies to develop high quality human capital in Saxony and the rest of East Germany? Are such subsidies the appropriate model for integrating Poland, the Czech Republic, Hungary and Slovenia in the EU? More on what's ahead for central Europe below.

Europe's big assignment

Was East Germany a 'suitable' candidate for the EU? No. Did West Germany follow an orthodox model of subsidies for the East German economy?

Yes. Did EU orthodoxy work in East Germany? No.

Is free mobility of goods, labor and capital within the EU (what the French label the 'pensée unique' or America's preference for market-based decisions and free enterprise capitalism) the answer? Perhaps. Or should East Asian approaches to governing the market prevail in central and eastern Europe? Perhaps. Is there a place for the two paradigms (that is, the market-rationale and the plan-rationale approach) in governing the market in the east–west borderlands? Perhaps. Is the west European reverence for social and market responsibilities within democratic states the most appropriate policy for central and eastern Europe? Perhaps.

Notice the tentative answers to the questions. Obviously, EU orthodoxy about subsidies has not created self-sustaining growth in East Germany. Some suggest that if they had it to do all over again they would not make these subsidies or, at least, not give DM900 billion (about $1.3 trillion in 1996 US dollars) to East Germany over the last seven years. Others suggest that it took West Germany almost 20 years after the end of World War II to create the German economic miracle, and East Germany needs more rather than less 'Marshall Plan' type aid or subsidies to recreate the German economic miracle.

Today, the Franco-German consortium prefers subsidies from Brussels and Bonn–Berlin over completely free market forces for East Germany. However, the successes in creating Asian-style economic growth in Poland, the Czech Republic and Hungary suggest reformers in those countries might be able to offer an alternative model to the current EU orthodoxy. The big assignment for Europe is to weigh the value of the alternative model, and to determine if and how it might fit within integration strategies chosen by the EU for Central Europe.

Win–lose win–lose world

The U-World is at the beginning of a different mix of trade and investment opportunities in Central Europe. Clearly, due to the local success stories, 'there is no longer a debate about economic theory. The fundamentals are not contested anymore,' says Rodric Braithwaite, a former UK ambassador to Moscow now with Deutsche Morgan Grenfell.[17] The hegemony of capitalism across central and eastern Europe is creating the potential for the rebirth of intraregional trade and investment, which collapsed along with the Soviet Bloc.

Poland and the Czech Republic grew by about 6% and Hungary grew by a little over 2% in 1996. All three are expected to continue their successes in 1997. Even with the expected slowdown in exports to the EU, these three Central European countries will increase their exports to their neighbors in eastern Europe, such as the Baltic states, Russia and Ukraine. In mid-1996, Poland joined the Organization for Economic Cooperation and Development (OECD), a group of 28 relatively wealthy nations which include the Czech Republic and Hungary. These decisions strengthen the opinion that the three countries of the Latin West are ready to join the EU by the year 2000.

'What needs to be done?' Rebuild industry with the most up-to-date technology possible. 'What do we get?' More steel, automobiles and trucks, and consumer goods. 'What can we do?' Secure partners from the West and invest in locally owned small and mid-size businesses. 'Tell me what you are going to do on Monday that's different?' Bet on a medium-term horizon for investments in Central Europe. Today, memory is best served by combining the desire for better goods and services with existing capital resources from the West.

Poland's emerging middle class

Naturally, if Poland keeps up its rate of economic growth, it will be like an Asian tiger, says Dan Lubash, head of European emerging-markets research at Merrill Lynch & Co.[18] However, Poland still has unemployment at about 15% nationwide, but it 'is down to about 5 percent in the big cities; Warsaw, for one, has more people at work than the similar-size German city of Hamburg ... The so-called hidden economy, too, is enormous. Lodz, hit hard by the downsizing of state-owned textile factories, has an official unemployment rate of 17 percent. But many workers start their own businesses, hiring former colleagues who work off the books.'[19]

Despite Poland's recent progress, per capita GDP remains around half that of Portugal and less than one-fourth Germany's output. Also Poland has resisted privatization and using other people's money to fix phones and pave highways.[20] Thus Telekomunikacja Polska SA, the Polish phone company, offers some of the continent's worst phone service. TPSA remains a sheltered monopolist even as other former communist countries have opened their telecommunications companies to foreigners.

Finally, about six million of Poland's 40 million people are middle class, 'puppies,' or Poland's yuppies who are buying cars, refrigerators, color televisions, washing machines, videocassette recorders, private day-care services, annual ski vacations to Austria and health care by private doctors.[21] Both GE Capital and Ford Credit, two big American providers of financing for cars, opened operations in Poland in 1995. According to Jerzy Kozinski, the marketing manager in Poland for General Motors, 'People's habits are changing ... they don't believe you have to have cash to buy. They know about credit and that with inflation dropping, interest rates are dropping ... If we define the middle class as an attitude, the figure is higher – more than 20 percent [or 8 million] and growing.'[22]

Given the cultural disposition in Poland to be middle class, 'more and more global companies are convinced they can earn a healthy return on their investments. Productivity in the region is rising fast enough to outweigh wage increases.'[23] In April 1996, Standard & Poor's Corp. rated Poland's debt investment-grade, a status the Czechs have enjoyed since 1993. 'There is now a first tier [of countries in Central Europe] for investors to target,' says Susanne Gahler, an economist at J.P. Morgan & Co.[24]

Percy Barnevik, the former chairman of ABB Asea Brown Boveri Ltd, a Swiss–Swedish industrial giant, put it this way: 'Poland was the first place in the region we could show that you can turn companies around'.[25] Poland is 'an inexpensive base to produce high-quality goods for pan-European sales. Central Europe is cheaper than the European Union and closer than Asia or Latin America.'[26] Poland's task is to shift its exports from low-end commodity goods to high-volume, high-value products, and to do this it needs secure markets within the EU.

German business habits in the Czech Republic

The Czech Republic took a different route to the glory land of capitalism.[27] First, it terminated its union with the more backward lands to the east, now an independent country called Slovakia, where Poland-A (the former Prussian–German lands) kept Poland-B (its more backward Asian lands on the border with Lithuania, Belarus and Ukraine). Second, the Czechs let large corporate interests (such as, German banks and firms, and American companies) take over the postprivatization initiatives from the government bureaucrats. Third, it let corporate interests buy and sell shares at will so that over one-half of the 1700 companies sold off by the state have a majority of private ownership.

Czech banks buy equity stakes in their biggest clients as the way in which to secure collateral.[28] This strategy is similar to the one used in Germany. Lender–owners nurse weak companies, cushion them from rapid change and force management changes. Without bank support on the board of directors, Czech firms are open to being taken over by foreign executives with a reputation for rapid restructurings. The latter has a rough-and-tumble business style that seeks to unlock the value of the assets.

VW acquires 70% of Skoda, introduces money saving German technology, employs Czechs at low wages and offers its Felicia at a low, low price of $8000 (in US dollars) – that is, $4000 less than Opel's Astra assembled in East Germany.[29] Right now the Czech Republic (as well as Poland and Hungary) is going through a car boom because everyone wants to replace their 20-year old junky, unreliable state-made Trabant and Skoda cars as soon as possible. The new cars from Skoda/Volkswagen in the Czech Republic and Slovakia, FSO/Daewoo and Fiat in Poland, and Opel/GM and Suzuki in Hungary are both cheap and reliable.[30]

According to Michael D. Dingman, the former president of Allied Signal Inc. and the owner of a smorgasbord of Czech companies, including Plzensky Prazdroj, a beer company and competitor against Budweiser, who bought his Czech portfolio long-distance from his beachfront home in the Bahamas from his neighbor Viktor Kozeny, the exiled founder of the Harvard Capital and Consulting Corporation, the Czech Republic's biggest investment fund: 'The historical education of Czech managers is different from American and Western managers – in marketing where you need to sell as opposed to allocate, and in the area of cost control'.[31] Dingman goes on the say this: Czech managers come from a classical German mold. They believe they own the market, and no will come in to shake up their territory. And 'they are frightened to death of [American] consumer companies who are so good and cunning'.[32]

Hungarians want American ambience

'Hungarians are known as trend-setters willing to latch onto anything new', says Sandor Demjan, the main Hungarian investor in the Polus Shopping Center Mall on the outskirts of Budapest. Hungarian entrepreneurs have put the following trends into Polus:[33]

- Cowboys and the American West.
- Palms and sunny Florida.
- Junk food heaven – McDonalds, Pizza Hut and Dunkin' Donuts.
- Movies with Hollywood action stars.
- American-style pricing and merchandising in retail stores.
- French-style hypermarkets selling food and clothing at low prices.
- Special offers on everything.
- Cheap imports from everywhere in the world.

As long as the prices are right, Hungarians will shop at the mall similar to Americans and west Europeans.

Also, as long as Hungary offers local talent to bankroll venture capital projects, Hungarians will get new shopping centers, new brand names and new investors. For example, Imre Somody set up Pharmavit to sell Plusssz vitamins and generic drugs. In 1995 he sold his firm to Bristol–Myers Squibb Co. Somody now is in charge of all Bristol–Myers products in Hungary, from cancer drugs to Clairol hair care.

Today, Hungary looks like a safe place for foreign investors. Germany's biggest bank, Deutsche Bank, is in Hungary for the long term. German exports to all three Central European countries are bigger than to the US, and where business goes, the banks go, too. 'We are relationship-oriented, not just deal oriented', says Hubert Pandza, head of the eastern Europe/Central Asia department at Deutsche bank.[34] 'We can help local companies with trade finance, corporate and project finance, mergers and acquisitions advice and treasury business', says Eugen Schuba, head of the east European department at Bayerische Vereinsbank.[35] Right now the business opportunities in central and eastern Europe cannot be estimated highly enough for all the German banks willing to set up shop in Hungary.

Due to foreign investment from German and other European firms, two-thirds of Hungarian exports to the EU are technology intensive, a higher share than for the three other Visegrad states – Poland, the Czech Republic and Slovakia – or Slovenia.[36] On the other hand, the proportion of Hungarian exports of sensitive commodity goods such as steel and chemicals is the lowest among the other Central European countries. Both facts bode well for Hungary's inclusion within an expanded EU.

What kind of 'suitable' members?

Although all Central European countries still have some way to go to fulfil Maastricht criteria and to close the gap in GDP with even the poorer EU members, the question is no longer whether they are suitable or whether they will join an enlarged EU but when, and what kind of members they will be.

The transformation to a market economy is all but complete. Poland holds back in privatizing the telecommunications sector. The Czech Republic pushed everything out of the state sector and let quick-study capitalists take the risks of transforming these enterprises into competitive firms. Hungary did the same. All have had some trouble putting in place the right kind of government oversight with the proper set of market-based regulations for banks and nonbank financial institutions.

All three are members of the OECD. Standard & Poor's rates their sovereign debt investment grade. Poland and the Czech Republic have growth rates equal to the best among the Asian tigers. Hungary has about half these rates, but well within the level to make its growth rate respectable among west European countries.

According to Tony Judt, a New York University historian, 'the EU, motivated in part by sentiment, would probably extend its "protective" arm around the centers of the Hapsburg empire: Poland, the Czech Republic, Hungary and perhaps Slovenia ... These countries will scrape into the fold of Europe ... the others will not. Thus, there will develop a sort of depressed Euro-suburb beyond which "Byzantine Europe" would be made to fend for itself, too close to Russia and Russian interest for it to be prudent for the West to make an aggressive show of absorption and engagement.'[37]

What kind of 'unsuitable' members?

The east European countries have a long way to go to close the gap in GDP with both the poorer EU and forthcoming 'suitable' members. The question remains how 'unsuitable' are Ukraine, Russia, Romania and Bulgaria for full membership in the EU.

Most of them have begun to liberalize prices and foreign exchange, carry out some privatizations and started on structural reforms in banking, transportation, communications, and budget and fiscal matters. Their transformation to a market economy is incomplete. All are still working on the proper set of government regulations for banks and enterprises. None have been admitted to the OECD. Their sovereign debt is not investment grade. And all have their own economic problems peculiar to a country's past, present-day transition and expectations about the future.

Ukraine: hostage to the past

'A third of Ukraine's exports consists of metals – mainly steel – and another third is chemicals, textiles and agricultural produce. These are all regarded as sensitive products by the EU, with imports subject to various quantitative restrictions and antidumping measures ... [Thus] half of Ukraine's trade is still with Russia'[38] rather than the west European countries. Recently, Ukraine has been trying vigorously trying to recast itself as a central European country. Although trade with Poland is only 2% of all Ukrainian foreign trade, exports to Poland are picking up, especially in cotton and the cross-border trade in consumer goods. Unfortunately, Ukraine's foreign trade is held hostage to the clash of elites from Dnepropetrovsk, and from the Donbass and Poltava regions (all in Russian-speaking east Ukraine) rather than Galacia in Ukrainian-speaking west Ukraine. In fact, Dnepropetrovsk got 10% of Ukraine's cumulative foreign investment since 1991 because of its 'connections' in the capital, Kiev. In short, Ukraine has been slow in cracking down on official corruption and unofficial personal influence between officials in government, state-owned industries and private interests.[39]

The result: in 1996, Ukraine's GDP fell about 8% and domestic investment fell by 27% of GDP. Foreign investment from Germany or the US is very low. Inflation runs amok and kills the chance of Ukraine developing a fully functioning market economy any time soon. Is Ukraine an 'unsuitable' member for the EU? Yes. Can it become a 'suitable' member any time soon? No. These answers do not come from the historical conflict between the Latin West and the Orthodox East, and the need to find a *rapprochement* in the Eurasian steppes. Instead, these answers come from a careful review of Ukraine's economic performance since 1991. Ukraine is still held hostage to its past within the former Soviet Union.

Russia as a potential competitor

Russians are not hostile to private enterprise. They want 'fair prices.' Since 1991 markets have taken root and Russians are just as keen as anyone else to make a profit. Also they want a better distribution of income. Where do they differ from Americans and west and central Europeans? If Russians are Slavophiles, they worry about the West's penchant for individualism, secularism, materialism and consumerism, and some even consider themselves anti-Western because Russia is a special place, once the 'Third Rome,' then a Czarist and Commissar Eurasian empire, and now a not so normal country at Europe's frontier with Asia.

Some argue that Russia's economy and standard of living are almost twice as high as officially reported. 'Russia's private sector, now constituting more than half of the economy, has been growing by 15 to 150 percent annually ... Furthermore, about 90 percent of private sector income and about 40 percent of all wages are never reported to Goskomstat and are therefore not reflected in the official reports of Russia's GDP ... This unreported earnings are part of the informal economy, an economy that by now may be larger than the formal economy.'[40]

The EU and the US should look at Russia as a potential industrial competitor. It could dominate some world markets in metals, oil and gas, chemicals and automobiles. The flow of these products to the West could upset prices, returns on investments and expected shareholder value within the international economy. By the year 2020 Russia 'may well have outstripped countries like Poland, Hungary, Brazil and Mexico with China far behind'.[41]

This is one future for Russia and its relationships with the West.

The result: the divergence between what has happened in Central Europe, especially in Poland, the Czech Republic and Hungary, and the former Soviet Union is striking. The European Bank for Reconstruction and Development concludes that GDP in Russia fell by 44% in 1996, and life expectancy for males dropped from 65 in 1989 to 57 in 1993.[42] Hyper-inflation has had a negative impact on economic growth and it has forced instability on the Russian economy. A careful review of Russian economic performance suggests this east European country, even if it wanted to, would not be a 'suitable' member for the EU. Russia along with almost all the countries of the former Soviet Union is held hostage by the past.

Delusional dogma

Given the right temptations, nations and peoples are capable of almost anything. Historians show us that the Latin West and the Orthodox and Moslem East fought over some of the lands in Poland-B (east Poland), Slovakia (the eastern portion of the former Czechoslovakia) and the Hungarian borderlands with Romania. From the beginnings of the Renaissance some 500 years ago, through the ups and downs of empires, wars and revolutionary communism, the Latin West won control over the peoples of the eastern marchlands:

- German forest tribes who became citizen-soldiers of the West Roman Empire, settlers in Hanseatic League cities of the Baltic Coast, the largest national population in Europe's center and the dominant force today in the EU.
- Polish peasant farmers who moved from the Tatra mountains towards the Baltic Coastal lands of the *ancien* Prussians and Wettins, got partitioned, were pushed west by the Russians, took over East German lands and seek permanent protection under the EU.
- Baltic fishermen who were overwhelmed first by the Germanized Prussians, Swedish and Poles, and then by the Russians, who now remain in three very small countries west of Russia and who wish to be saved from Russia by the EU.

- Carpatho-Ruthenians and Galician Poles who were attached to different empires and nation-states as borders shifted, and whose land now belongs to Ukraine.
- Bohemian merchants who became prosperous under the Hapsburgs, lost everything under the Nazi and Communist occupations, and now seek rebirth within the EU.
- Hungarian horse breeders who pushed back the Turks from Vienna, later shared the Hapsburg crown with the Austrians, lost vast territories to Slovakia and Romania in the two World Wars, and now seek renewal within the EU.
- Croatians who as south Slavs preferred Roman Catholicism, the Latin alphabet, the Hapsburgs and the German *lingua franca* to the other fanatacisms of greater Serbia, a southern Slav kingdom and communist state, and the former Yugoslavia Slovenians who are not Italians, Germans or Slavs, and who want to be a part of the EU for trade and investment reasons.

Today, the global, largely American consumer life style is working through the EU and multinational enterprises to move the commercial boundaries of the West once again to the east. Without letting the east European countries into the EU, without this political settlement, the West cannot move eastward towards the Eurasia steppes. Both Europe and America might suffer in the distant future from the EU's current unwillingness to be flexible towards the aspirations of all the nation-states in eastern Europe.

Imagine a different history still, with east Europeans as the founding fathers of a free trade agreement for all of Europe. The entrance requirements would be just as divisional, given the desire to declare nearby countries in the Orthodox and Moslem East as 'suitable' members, and countries in the Latin West (especially those along the Atlantic Coast) as 'unsuitable' potential members. Unfortunately, uncritical worship of the West Roman Catholic–Protestant past has flourished from the time west Europeans took to their boats and conquered the Old and New Worlds. As a result, west Europeans now disguise Europe's Mediterranean and Eurasian steppe past with a 'delusional dogma' that tends toward claims of cultural superiority.

The Orthodox and Moslem East fail to get into the EU because their recent economic history has not been as successful as that in the Latin West. They fail to get in because the Latin West shunts them aside as historical relics – a problem that is particularly nasty towards Russia, Ukraine, Romania, Bulgaria and Turkey.

Keeping these east Europeans out of the EU only isolates them more. Also the west Europeans recreate the old problem of Germany – how to keep the marchlands east of somewhere to somewhere further inside Eurasia from exploding with rage, fire and storm. Europe in the east has no natural frontiers and needs the cement of free trade and investment to keep it working in concert with western Europe, the EU and the US.

Euro-realism

Here we have Euro-realism in a nutshell. Inventing Europe means going beyond the old thought processes about what is and what isn't Europe. Adding all the central and east European countries that want in to the EU is the best chance for making all European nation-states suitable partners in the quest for Europe's future. The suitability of Poland, the Czech Republic, Hungary and Slovenia is unquestioned on historical, cultural, economic and political terms. The West asks this question: are they or aren't they? The answer is yes they are Europeans.

The suitability of the Baltic states and Ukraine falters on the prospective negative Russian response; the West believes it gets very little of value by admitting these east European states over the strong objections of Russia. Thus Ukraine, which has a population as large as France, will be left to its own devices rather than as a force for Western influence in Galicia and the capitol region surrounding Kiev. The West asks this question: are they or aren't they? The answer is yes they are Europeans, but they are not the right kind of Europeans for the EU club.

The unsuitability of Romania, Bulgaria, the states from the former Yugoslavia and the former Soviet Union, Turkey and Russia is unquestioned on cultural, economic and political terms. The West does not bother to ask the question for it knows for sure the answer will be no.

Inventing Europe means tempering our ideas of 'are they or aren't they' with the realities of the *Wende*, the hegemony of money capitalism, emerging middle classes, German business habits and American ambiance. We must give up the delusional dogmas we carry around because of the corrosive effect of orthodoxies either from the East or the West. Understanding this crucial idea is the first step towards accepting Euro-realism.

Acquis communautaire

If EU leaders can get beyond their corrosive orthodoxies, they might even develop a common vision for Europe. This means a reduction in the cultural bickering among the Latin West, the Orthodox East and Moslem south-east; a coming together of all three in a common economic effort to expand trade and investment within Europe; and a joint effort to build up Europe as America's partner in the North Atlantic. Let's try our hand in writing a vision for (western, central and eastern) Europe.

First, the EU must survive but not as it is today. The EU must give up its unified approach to integration and adopt a two-step approach to Europe's future. This goes by several other names including flexibility, variable geometry and multispeed approach.

Second, the current members of the EU that are willing and able to integrate fully should do so in trade, investment, monetary, labor and capital matters. Those that are willing but unable to join the EMU or the Social Chapter should be allowed to opt out with no penalty.

Third, new members of the EU should be given a choice of full integration or flexible integration. The choice is up to the new members as they negotiate their admission to the EU, but their accession cannot harm the Franco-German consortium that runs the EU.

Finally, both current and new members must remain subject to the *acquis communautaire*:[43] the single market; free trade in goods, services, labor and capital; and price competition. Also both current and new members should accept the two-stage approach towards the EMU, and the accession of new members from central and eastern Europe. Moreover, current and new members should be able to opt out of the EMU and other EU programs without damaging the economies of those members that go forward with these programs. (This is a lesson learned from the competitive devaluation pursued by the UK after its expulsion from the ERM in 1992.)

In summary, the EMU is *not* the one crucial element in the further integration of Europe for the twenty-first century. The claim of the Franco-German consortium that the EMU is the single most important integration issue facing the EU rings hollow; the EMU is crucial only to a core of countries including France, Germany and the Benelux countries; the EMU is useful but not important to all the rest of the countries within the EU. However, the accession of new members from central and eastern Europe is the one crucial element in the long-term success of the EU and its future partnership with the US.

Here is how Valléry Giscard D'Estaing, former French President, puts it: 'If the European Community has succeeded in surviving and, even, in making some progress, this has always been at the price of maintaining a persistent ambiguity as to its ultimate destination'.[44] This seems evident in the struggle over the EMU, the accession of new members and the *acquis communautaire.*

Costs and benefits

How should Europe weigh the costs of and benefits of adding new members to the EU?

This is a case of decision-making under uncertainty. If only the three countries from Central Europe get asked to join, the member states will feel the task can be done without too much risk to their fiscal solvency. On the other hand, if all who want to join are asked to become members, the EU will be unable to accommodate them within the present budgetary and decision-making framework now in place among the member-states. How many new member states beyond the three could be asked to join without overstressing the EU is unknown.

Decision theory cannot create certainty where none exists, but it offers a way to organize one's thoughts and gather the information needed to make better decisions. Consider adding new members. The best policy is to do it quickly and get the three Central European states in as rapidly as possible.

Remember Europe has choices. One option is to broaden the EU through a two-step approach towards membership in the EU. The three central European countries enter first. Then others enter later. Each accession vote will be difficult within the EU, among the member states and in the candidate country. When all is said and done, both the candidates and the existing EU members might lose patience with the admissions process. Some east European countries might decide not to join (such as Russia), or they might be rejected outright by the EU (such as Ukraine and Turkey), or they might be put on hold by the EU for an indefinite time in the future (such as Romania and Bulgaria). The whole process might end in failure for one or more east European countries.

A second option is to go for broke and do everything possible to bring in all the members who want in. Nobody thinks the EU can pull this off, but the sheer audacity of it might turn heads in western Europe. It would be a shame if the EU stumbles because no one was willing to work on bringing the whole of Europe together.

Today, the EU is pursuing option one without regard to the true downside costs to the EU. Will the member states continue to be happy with this state of affairs? Yes. Because so many national leaders have staked their political reputations on opening the door to the three central European states. However, they should be ready for some unhappy campers in the Baltic states, Ukraine, Romania, Bulgaria and Turkey. The EU should be ready to offer these 'unsuitable' countries another alternative, one that is appropriate to their current economic status. This type of enlargement fits into the concept of 'variable geometry' for Europe.

Decision theory offers us a way to reflect on our choices and make the best decision we can with the information we have in front of us. Decision theory offers Europe a way to organize itself to beat the fate of Salmon Day.

Salmon Day

Salmon Day is the experience of spending an entire day swimming upstream only to get screwed in the end. All foreign investors in central Europe seek market share without a certainty that the three countries will become members of the EU. These investors are betting on the come. Some of these foreign investment deals in central Europe might fall victim to Salmon Day.

Here are some summary thoughts. First, the transmitters of Europe's common knowledge about the big idea called the European Union offer the following: participation in an expanding free trade area; relief from petty squabbles over tariffs, customs procedures and non-tariff barriers; and help in making Europe whole once again.

Second, the editors of Europe's common knowledge are in the process of reviewing the start-up costs of three or more new members to the EU, and speculating on how a two-step approach might avoid the cost of aid and subsidies to the new member-states. Some of their vivid memories includes the West German expense at taking over the East German economy.

Third, managers and others in the professional class must decide whether the addition of new members is more important than the single currency. Inventing Europe depends on adding new member-states with some degree of flexibility (or variable geometry to use Euro-speak). If some of the new members exchange their right to aid and subsidies for the free movement of their labor, this might be sufficient inducement for the member-states to admit the new east European members. Here then is Europe's danger as it swims upstream with the probability of getting screwed in the end.

Wisdom: decision theory lays out possible choices. Pay attention: one choice leads to Salmon Day. The other might avoid Salmon Day. Open the doors: Europe makes a choice to go forward by admitting a limited number of central and east European countries that want to join the EU.

What do I get?

Here is our to do list:

- Buy into Europe's big assignment for adding new members.
- Accept that only some potential members will be admitted.
- Build on the memories of deliberate gloom in east Germany.
- Reimage ideas about a two-step admissions policy.

- Recognize good infrastructure attracts foreign investors.
- Create more trade in goods and services with the East.
- Use decision theory to weigh costs and benefits of options.

Remember these words. Salmon Day for the EU can be avoided, but the Franco-German consortium will have to rethink its approach towards admitting new members to be sure Salmon Day is avoided for the EU.

Successful decisions

A successful Europe makes the following policy decisions:

- Europe works with America to create a North Atlantic Free Trade Area.
- Europe works in the east–west borderlands to create free trade in the East.
- Europe and America let the WTO help Orthodox and Moslem communities in the East.
- Europe gives limited assistance to Russia, Ukraine and Central Asia.

Part 3

East Asia Discovers the Way

How to reimage Asia's place in the global economy

Japan Stalled
China Realism
Paradigms for South East Asia

Executive decisions

Remember East Asia has choices. In Chapters 7 through 9, the following choices are discussed. They are divided into two categories. The first set has a high probability of success in avoiding Salmon Day for East Asia. The second set has a low probability of success in dodging Salmon Day for East Asia. You decide what to do.

Decisions leading to success

- Japan reforms itself at home and becomes a stronger competitor abroad.
- Japan opts for free trade with the US and freer trade within the Asia–Pacific region.
- China accelerates reform of its socialist system with market characteristics.
- South East Asia pursues gradual and harmonious reform of free trade within Asean.
- South East Asia accelerates reform by linking up Asean and APEC free trade initiatives.

Decisions leading to failure

- Japan practices a mastery of drift and alienates both the US and China.
- Japan opts for closer ties with China and these alienate the US and South East Asia.
- China goes too fast on the path of reform.
- South East Asia staples itself to either Japan or China or both, and loses the US.
- South East Asia accelerates reform by linking up Asean and NAFTA free trade initiatives.

The work of Salmon Day is about making decisions under uncertainty.

Chapter 7

Japan Stalled

'As the world's second-largest economy, Japan boosts per capita purchasing power almost twice that of Hong Kong and 80 times that of China.'[1]

Robert Neff, *Business Week.*

'What may be surprising is that Japanese growth seems to be stalled …, at a point where output per worker is well below American levels.[2]

Milton Ezrati, Chief Investment Officer,
Nomura Capital Management, Inc.

Memories always win

Divining what connects Japan and China and what each means to the other is not within the memory of most Americans and Europeans.

Westerners see Japan in Japanese terms – the second most important industrial nation-state among the G3; an economy in which government and business close ranks to solve problems; and a copycat consumer society for luxury goods, fast foods and nostalgia trivia. Japanese cultural practices are mysterious to most folks in the US and Europe.

Americans and Europeans see China in Japanese terms – a third world nation-state with many first world talents in its coastal regions; and an economy in which market forces are balanced (some say burdened) by socialist practices. Chinese cultural traditions are mysterious to most westerners, but not to the Japanese, Koreans and others in East Asia.

Both Japan and China prefer the plan-rationale approach towards trade and investment over the market-rationale approach towards the international economy. In this they are alike. However, Japan and China differ over their collective future. China wants the West to see Japan in Chinese terms. Japan has yet to decide whether to defer to China's new found boldness. Frankly, most folks in the West do not know this competition for the hearts and minds of Americans and Europeans is going on between Japan and China.

Are we in the West uninformed because we have a fascination with the ancient sages of the Orient? Are we in the West misinformed about the mixture of respect, disdain, emulation, and rivalry between China and Japan? Are we in the West confused about the role *kabuki* plays in Japan's relationship with China, the US and Europe? Are we in the West being gulled into thinking that the plan-rationale approach (or the Confucian version of central planning) is not a huge barrier to achieving the material quality of life taken for granted in the affluent US and Europe? The answer is 'Yes'.

Japanese memory of 1000 years of Chinese influence includes the following:[3]

- Chinese characters form the written language of Japan.
- Confucianism and Buddhism shape Japan's religious and cultural sensibility.
- Cosmology about Emperor-worship connects the natural and human worlds.
- Culture encourages the practice of Zen aestheticism and the writing of *haiku* poetry.

These four imprints forged Japan's Sinicization.

Political memory

With rapid modernization some in Japan pushed the country to forget its Chinese past, leave Asia completely and cleave to the West. Others suggested Japan lead Asia against the West. Over the last 100 years, Japan both understood and misunderstood China's willingness to let Japan take the leading role throughout East Asia.

For example, during the pre-World War I period, when China's Last Emperor neither reigned nor ruled, Japan's early successes in countering the West offered political and economic models for China's nationalist revolutionaries under Dr Sun Yat-Sen. In the interwar period, a more aggressive Japan left its language, and built a substantial industrial and transportation infrastructure in Taiwan and Manchuria, respectively. During World War II, Japan failed to organize Asia against the West, and instead drew the US fully into Asia on the side of China.

Within a few years after this war, the tables were turned. Japan allied itself with the US against China's communist revolutionaries. In due course Japan became an economic powerhouse in the global, largely American economy. On the other hand, China languished in ancient poverty and modern Leninist–Maoist statism.

Today, China is again following Japan's example by exporting large volumes of goods into the world's market economy. Also Japanese, American and European firms are making substantial direct investments in Manchuria, Shanghai and southern China. Moreover, a consumer market is being built in coastal China with the same interest in comparing foreign and local brands for value, price and quality that is found in Japan and throughout East Asia.

All of this is being done with the memory of Japan's past mistakes in handling its political relationship with China. As Japan was in the 1930s and 1940s, China will be a contender for the world's food, oil and minerals. 'One fairly cautious Chinese scientist says [China] will need to import 33m tons of food a year by 2020. As for oil, Kent Calder of the Centre for Strategic and International Studies in Washington, DC, thinks that by 2010 the Chinese will each year be wanting to bring in form outside the equivalent of half of Saudi Arabia's total present production.'[4] Of course, if Japan becomes too Western, then China and other neighbors unite against it and constrain its ability to buy food, oil and minerals from abroad. If Japan remains Asian – that is, knowing its place and deferring to China – then Japan prospers, helps its neighbors and enjoys their collective support in the quest for world markets. A consequence of a Japanese decision to opt for Asia and China is as follows: Japanese trade and investment will be drawn away from the West and channeled toward Asia, especially China.

Kabuki *memory*

Japanese bureaucrats and business executives are carrying out an incremental reinterpretation of their plan-rationale system of economic governance. Between 1991 and 1996, Japan had an almost motionless GDP, or five years of near zero economic growth. Instead, the national economy went through asset deflation and recession.

These key players in government and business are a loose coalition of semiautonomous groups with fierce communal loyalty, a strong commitment to their place of work and a slow process of decision making. Community forms the overarching principles of social organization in Japan. To move from a plan-rationale to a market-rationale (in which markets form the overarching principles of social organization) these core decision makers need a revolution in consciousness, one that moves them away from directed economic activities in order to achieve predetermined outcomes.

None of this is going to happen because the Japanese economy remains relatively closed. Japan is still a fortress. Compare these numbers:[5]

- The US accounts for 22% of world economic output and absorbs 21% of global investments from Canada, Japan, the United Kingdom and Europe. Foreign direct investments in the US equals $68 billion (in US dollars).
- The EU accounts for 19% of global GDP and takes in 39% of foreign direct investments from the US and Japan.
- Japan accounts for 15% of world economic output an absorbs only 1% of global investments from the US and Europe. Foreign direct investments in Japan equals $3.1 billion (in US dollars).

However, things might be changing just a little bit. Imports of personal computers, cellphones and semiconductors continue to grow, and more foreign-trained Japanese are reaching top executive ranks. And 'overall foreign investment in Japan may have jumped more than 50% in 1996 ... The [mergers and acquisitions] deals involve big names, too – food-processing by Dole Food, Glaxo Welcome's full acquisition of its local joint venture, and moves by GE capital to get into consumer finance.'[6]

Moreover, GDP began to grow again in the fiscal year ending March 1997. After all, Japan is 'the world's second-largest economy, seven times the size of trendy China's. Japan's GDP makes up a whopping 73% of the total gross domestic product of all the countries in East Asia.[7] Lastly, 'as the world's second-largest economy, Japan boosts per capita purchasing power almost twice that of Hong Kong and 80 times that of China'.[8]

Nevertheless, economic activity in Japan and East Asia is an instrument of national power. That's why it is important to pay attention to the steps currently being taken by Japan to protect its own interests in the future. Japan is eager to accelerate growth and help industries restructure within its plan-rationale approach towards the domestic economy, and its commitments to the China, the other nations of East Asia and the US.

Japan's direct investment in China reflects the collective Japanese recognition that China is on the march towards long-term, noninflationary economic growth at rates of 8% per year. One half of the GDP goes into building a consumer society within China, and the other half flows into building up the Peoples' Liberation Army (PLA) and other Chinese military forces.

Here is the forthcoming challenge to and confrontation of Japan. If the US is unwilling or unable to help Japan with its future China problem, the US will loose its military partner and chief foreign source of financial support for the dollar. Japan's Ministry of Finance (MOF) and Ministry of International Trade and Investment (MITI) will look for a way out – that is deferring to China, the PLA and the plan-rationale forces within the Chinese bureaucracy.

Tomorrow in the western Pacific is unpredictable. If the US–Japan partnership continues, both market-rationale and plan-rationale will be the two competing overarching principles of social organization between the two countries and among the countries of North Asia and South East Asia. If the China–Japan partnership gains strength, only plan-rationale will be the overarching principle of social organization for all countries in East

Asia, India, the Muslim countries of South and Central Asia, and, perhaps, even Russia in the Orthodox East.

With China as a partner Japan will become a higher-tech Canada as it works out its Canadaization by the giant next door, China. Rules of feudal loyalty from Confucianism, Buddhism and Shintoism demand Japan push China's admission into the World Trade Organization, and link both countries in a supranational trading partnership.

Does Japan share the broad aims of the US or does it cover its intentions with white facial powder? If Japan says it does share these goals, will it fight for them or simply pay lip service to these pan-Pacific aims? Can the US comprehend Japanese decision making concerning overarching ways to achieve social goals?

We in the West won't know whether our correct answers are 'Yes', 'No' or 'Maybe'. Periodically, Japanese bureaucrats and business executives go through the process of reinterpreting their socio-economic and political life. This habit goes back a thousand years. Incremental reinterpretation forms the basis of *kabuki* – that is, an ongoing play in which the parts are acted out so that no two people interpret the drama in exactly the same manner. *Kabuki* is a way to keep Chinese, Americans, Europeans and all foreigners from controlling Japanese political, economic and social institutions.

Loyalists versus U-World

Our analysis from Parts 1 and 2 draws the following conclusions. As business executives seek to bridge the cultural divide between East and West, the cultural facts within the European Union, between North America and Europe, and among the Latin West, Orthodox East and Moslem lands become more different in both detail and substance. The facts of global materialism (such as, Coke, Big Mac, Cadburys, BMW and Walkman) shape the loyalists view of the U-World and determine whether the former wants to join the rush towards globalization, multinationalism and internationalism.

Here is where we are going in Part 3. Chinese, Japanese, Koreans and others in East Asia are born into societies in which they recognize wide cultural differences from their commitment to Confucianism, Buddhism and Shintoism. These beliefs give them overarching principles for social organization different to those in the world economy dominated by the West. Although a generation of Japanese has grown up with relatively free access to US markets, it has done so through a plan-rationale approach towards the Japanese domestic economy and acceptance of a market-rationale approach towards the West's international economy.

Within East Asia many national communities learn early to defer to China. Japan is no exception. Deference to China's social habits and commitment to US economic values help keep Japan free of competing claims on its national institutions. Long ago the Japanese thought they had conquered China, but they had not. In fact, they woke up to discover they too are 'Chinese' in belief, mood and custom.

Cultural appearances

Fortress Japan means to US observers that bureaucrats run an economy in which the producer comes first and the consumer a distant second. To their untrained eye Japan is going through rapid change, a deep transformation and a reimaged culture. They point to these examples.[9]

First, as Japan struggles to meet global competition, Japanese employees face job insecurity, economic dislocation, higher levels of unemployment and an ageing population. Some large industrial firms are cutting their workforce through early retirement and moving personnel from cost to profit centers. With Japan's record-high jobless rate of 3.4% a small amount of wage competition is developing in labor markets.

Second, these firms are doing more sourcing offshore, and disrupting traditional loyalties and distribution arrangements. A larger amount of foreign-sourced components and finished goods (such as Asian-made color TV sets, and US software and personal computers) are driving domestic suppliers into bankruptcy.

Third, domestic retail prices for men's suits, cars, computer chips are coming down as younger Japanese learn to vote with their wallets for the newer discount retailers, or use the fax to buy from US mail-order companies. A small amount of price competition is developing in retail markets, and this seems to be picking up steam throughout the urban areas of Japan.

Finally, the bubble economy sent land prices soaring and then they come crashing down, wiping out asset values in the stock market, Jusen housing loan companies, bank reserves and the credibility of the MOF.

None of these in themselves are rapid changes. Together they do not point to a deeper transformation. The conclusion is inescapable: this is not cultural reimaging. The U-World is not winning out over Japanese loyalists. Rather, the latter are trying to decide whether they will be rooted cosmopolitans – namely, disseminators of free trade ideas, upper middle-class spenders rather than savers, promoters of the computer-based digital revolution, and deal-makers for East Asia and the entire pan-Pacific community.

Unwinding red tape

Shoichiro Toyoda, the chairman of the Toyota Motor Company and of Keidanren, the Japan Federation of Economic Organizations, recommends the following bold reforms:

(1) Complete deregulation of the Japanese economy to lower the cost of doing business and to promote competition in Japan.[10] Note: 'Electricity prices are about a third above those in America, road transport costs about 40% higher, and coastal shipping and international telephone charges three times as expensive.'[11]

(2) Lower taxes to create new jobs and to rebuild the industrial base in Japan.[12] Note: 'Output per worker in Japan, which grew at an annual pace of 6–8 percent for much of the century, began to decelerate in the early 1970s … to 4 percent … in the mid-1980s, and most recently, to just 3.5 percent … What may be surprising is that Japanese growth seems to have stalled …, at a point where output per worker is well below American levels … [Also] average return on new investments in Japan is well below the return [in the US].'[13] Without a rapid solution to these ongoing economic problems, Milton

Ezrati the chief investment officer at Nomura Capital Management suggests Japan's power in East Asia may ebb while China's power might gather strength together with or in conflict with the US.[14]

Prosperous East Asians

Japan is indeed going through an incremental reinterpretation of the way it relates to competition from other first-world, developed countries, such as the US and Europe; and how it integrates its continuing commitments to China and East Asia. A great deal of what changes and what stays the same depends on whether the Japanese middle class returns to the conspicuous consumption of the 1980s. Historically, they save 13% of their annual income versus 5% in the US. Consumers account for 60% of the Japanese economy, and only their increased spending can boost the Japanese economy into full recovery.

Unfortunately, given the fear of job losses among the middle class in Japan their urge to splurge is down for the sixth year in a row. This is not the case in China. 'For the country's growing class of the newly rich … a ludicrously high price tag is all part of the package. In recent years, China's conspicuous consumers have made the purchase of over-priced cognacs one of their favourite ways of flaunting wealth, in some cases buying bottles priced at several hundred dollars only to smash them on the floor.'[15] Well-heeled expatriates among the American, European and Japanese communities in China would not do this to good wine, brandy and cognac.

Today in 1997 prosperity affects East Asians differently. Will Japanese rooted cosmopolitans continue to save at high levels and defer instant gratification? Or will their commitment to community values encourage them to boost spending and hence help the Japanese economy into full recovery? Will the very prosperous Chinese continue to spend a lot more than all other segments of the Chinese population? Will a new wave of tightening tax laws and consumer spending force them into reducing their conspicuous consumption?

On these issues, Japan and China will come up with different answers. Japan and Singapore are essentially middle-class countries in which ostentatious consumption is frowned upon except when business executives get together to play golf or meet socially in nonbusiness settings. On the other hand, China is not a middle-class country, and it has even wider extremes of income than are visible in Thailand, The Philippines, Malaysia and Indonesia. All of these countries have large communities of overseas Chinese who are richer and wealthier than their conationals from the Thai and Malay communities. Yet all these countries seek broad-based economic growth with a widening middle class who saves and spends along the lines of the Japanese.

There is the paradox of economic life in East Asia. Deference to China in social and political things is essential among the nations in the region. However, deference stops when it comes to economics. The plan-rationale approach of Japan finds favor in Korea, South East Asia and even in China and Vietnam, the two countries with a significant Marxist–Leninist present. The market-rationale approach of the US tends to dominate thinking in The Philippines, and is paid lip service to elsewhere in the region, especially in bilateral dealings with US itself and in their collective response to the WTO.

East Asia has become prosperous by being able to chose the best from both worlds. This is the vivid memory of the past few decades. It will continue for the foreseeable future.

Win–win world

The U-World is at the end of the beginning of a new win–win world market era for Japanese firms. Here are the crucial questions for Japanese CEOs:

'What do we get?' New joint ventures and alliances with US and European firms. New knowledge, better information, more sales and improved human resources management. Application of cultural assets to real business problems. The corporate leaders have kept Japanese factories open in the face of high-wage stiff competition from the US and Europe, and low-wage competition from China and the countries of South East Asia.

'What can we do?' Be a prophet about globalization. Forecast correctly

about long-term changes in important industries. Switch to overseas suppliers. Lay off managers and workers. End public life pathologies between business and political leaders. The CEOs are taking their firms through a revolution in thoughts, words and deeds.

'Tell me what you are going to do on Monday that's different.' Let's look at several examples.

Softbank

The cyber-mogul, Masayoshi Son, the grandson of Korean immigrants to and a member of a despised minority in Japan, is an example of how to straddle several worlds and come out a success. He is, perhaps, the fourth richest person in Japan. The few wealthier men probably inherited their real estate fortunes from parents. Son is one of the few self-made men in Japan. 'After a $3 billion, two-year, debt-fueled acquisition tear, Son, the president and CEO of Softbank, already has emerged as the world's biggest computer trade-show impresario and most powerful publisher of high-tech magazines [Some were bought from Ziff-Davis Publishing]. Also Softbank is pumping big money into unproven but promising Internet ventures,'[16] such as search-engine maker Yahoo.

In 1996, Son and Rupert Murdoch jumped into the high-tech world of digital broadcasting and took a 21.4% share in TV Asahi; their intention was to secure programming for BskyB, Murdoch's digital TV venture that aims to offer 150 channels to Japanese viewers. According to Yutaka Sugiyama, senior analyst at UBS Securities, 'Culturally speaking it's quite difficult … for foreigners [to barge] their way into … a key business as television … There's a strong resistance in Japan.'[17] So within nine months Japan's media powers closed the business to outsiders, and Asahi Shimbun, which publishes a newspaper with 8 million subscribers, bought out the partnership. Neither Son or Murdoch lost money. They continued their prowl for a way to build a strong market position in Japan's new digital-TV business, and in early 1997 they signed a deal with Sony and Fuji to bring satellite digital-TV business to Japan.

As Son disrupts the old order of lumbering, tradition-bound electronics conglomerates in Japan, distinctions blur among Chinese, Korean, Japanese and US ideas. His cross-cultural bets include Web services/advertising, online publishing, electronic commerce, games and Web software. 'We have a healthy core market that is growing 20% to 30% year-on-year,'[18] says Son.

Kyocera

Another Japanese business executive who made good is Kazuo Inamori, the recently retired head of Kyocera, the $6 billion company that controls about 65% of the world market for ceramic casings that insulate and protect computer chips. He has worked very hard making 'Japan's closed economic system more receptive to iconoclasts like himself,'[19] Son Akio Morita, Sony's cofounder, and Soichiro Honda. 'Much like Sony and Honda, Kyocera had to prove itself first in the American market because Japanese companies were reluctant to buy from an unproven supplier.'[20] During the days of the bubble economy, Kyocera is one of the few Japanese companies that bought an American manufacturer of capacitors, AIX Corporation of South Carolina, and turned it into a success. Kyocera's other American acquisitions, such as the connector manufacturer, Elco Corporation, did not do as well.

Inamori is of the school of Japanese business executives that believes values can be taught in the factories and offices of companies. Japanese employees come to work early each morning to study his teachings about hard work and fanatical devotion to the company and leader. On the other hand, American employees do these early morning studies once or twice a week because to do more seems too onerous to American workers. Now retired Inamori practices his philosophy of material and spiritual happiness as a Zen Buddhist monk in Kyoto, Japan.

Japanese cultural assets

Many Japanese executives practice Zen and write *haiku*, or short poems of three lines of five, seven and five syllables that do not rhyme. About 5–10 million people in Japan write poetry regularly, and together with Emperor

Akihito and his family they enter their poems in the annual imperial poetry contest.[21] These poems are used to sell all types of products in Japan. Both Zen masters and *haiku* poets have celebrity status in Japan, a significant difference in how American and European executives carry out their managerial responsibilities.

Here's Japan at its best. Its business executives pass on the memories of Japanese culture and also engage in a few alternative futures. These leaders show the Japanese and others in East Asia how to swim along the currents of the U-World and still be crucial players within the loyalist nation-states. They offer China, and South East, South and Central Asia a choice.

Japan as loyalist nation-state

Some distinctions blur. What used to be were Japanese business executives who had an intimate knowledge of and faith in the company's products; close, almost familial membership of domestic management club; and were Japanese engineers. What is coming into place are Japanese chief executive officers who put a greater emphasis on sales and marketing, and who believe in globalization.

The work of Sony's prophet, Akio Morita, in global markets paved the way for what followed. For example, after five years in advertising, marketing and communications, Nobuyuki Idei became president of Sony. Also Minoru Makihara, who is London-born and Harvard-educated, became head of Mitsubishi Corporation. Moreover, Taizo Nishimuro, who built up Toshiba's multimedia businesses, is the first nonengineer to take over as president of Toshiba, a Japanese company whose core business is still heavy electrical industry.

Finally, Hiroshi Okuda who pioneered Toyota's US expansion is the first sales expert to take over the firm. Early in 1997 he surprised a Tokyo business luncheon by saying 'the auto industry is not immortal ... There hasn't been a single industry that continued to be profitable for more than 40–60 years.'[22] His plans include expanding Toyota's nonvehicle-manufacturing businesses, especially telecommunications, to about 10% of sales in 2000, and using start-up ventures to put more electronics in automobiles, voice-navigation systems and automatic toll booths. Here is where the real value-for-money is in the 'automobile' industry. By such diversification, Okuda hopes to overcome the slowdown in unit vehicle sales in Asia; they are a direct results of the traffic-chocked streets in Japan and the emerging markets in East Asia. Finally, Okuda needs to protect Toyota from its loss of market share in Japan.

This new breed of Japanese business executives does not want the continuation of the public life pathologies discussed in Chapter 1, such as the money payments from Japanese firms to LDP political leaders. Nor does this new breed want the enduring *keiretsu* cross-holding ties among families of business firms and MITI, or Japan Inc. Right now the biggest six *keiretsu* produce more than one-fifth of Japan's GDP. Many still co-ordinate strategy and deal predominately with each other.

Moreover, employees of money-losing firms within *keiretsu* get the same wages as workers in profitable units of the same business group.[23] None of this sits well with the new breed of Japanese chief executive officers because such high fixed costs cannot continue in the wake of more competitive international prices from American firms. During the recession that followed falling assets prices in the early 1990s, 'Unemployment rose to record highs ... But the pain was not evenly ... distributed. Younger people ... were sacrificed ... by older, inefficient and highly paid managers.'[24] Bloated middle management was not reduced. Instead, cheaper parts and components were brought in from overseas. The switch to offshore suppliers was a one-time event that is unavailable for the next recession. Japanese companies need productivity gains to match international standards. This means cutting an inefficient managerial staff and replacing it with new ideas, new people and new technologies.

Which vision of Japan will swim the ocean's currents without climbing the rivers to inevitable death? Does Salmon Day come to all nations? Or are there ways of escaping fate?

One-time events

The U-World forced two one-time events on Japan. These are as follows:

First, the need to restructure politics, finance and economics. This is being discussed within Japan. Some new approaches are being introduced, such as change in the way political parties are financed, higher sales taxes, limited reforms at the MOF, more openness in telecommunications, fewer cross-holding ties, etc. All of these are designed to make Japan more open to world business interests.

Second, the globalization of Japanese firms' markets, now less centered on Japan itself, and the acceleration of technological change. Here's what Makihara of Mitsubishi says: 'Technologies are now moving so fast that it is impossible for the top manager to know all details. Companies are now looking for generalists who can understand the broad changes, delegate and provide leadership.'[25]

Nishimuro of Toshiba is a straight-talking Confucian scholar who suggests his appointment as chief executive officer 'does not mark a complete break with the past, but rather a gentle shift to new priorities ... The main thrust ... is a shift in emphasis from product development, formerly at the heart of Toshiba ..., to marketing ... Product development has to be led by marketing ... because markets are taking the lead.'[26] With these changes will come layoffs among deputy department heads and section chiefs so that communication will be speeded up with top management, perhaps, even through E-mail – which is a rarity in Japanese companies.

Let us be careful. The most likely outcome is as follows: Japan is on the verge of going from something to something else. The Japanese government, its business firms and executives are slowly making some changes in how macro economic and micro business activities are carried out within Japan. The speed of these changes is less essential than their deepening in top, middle and lower management throughout the public and private sectors of Japanese society. If this deepening continues unabated for a decade or more, Salmon Day will be avoided. This is the big assignment for loyalists and one that might foster Japan's role as a leader in East Asia.

Paying attention to free trade

Today in Japan we have the ongoing redefinition of business strategy by Japanese firms. They are trying to remain domestic firms with strong ties to bureaucracy and *keiretsu*, and at the same time compete against foreign-owned international firms in terms of costs, productivity (or output per worker) and efficiency. Japanese are in the process of selecting those managerial habits that suit them best. Many of these new ideas about sales and marketing come from the US. The cultural evolution of the cosmopolitan elite within Japan is a positive force to economic growth on both sides of the North Pacific. These international managers have grown familiar with how the Americans and Europeans deal with free trade under NAFTA and the EU. Japanese managers believe the interlocking relationships among Japanese, Asean and Chinese firms are a plus; because these trade and investment deals push a growing interest in free trade among East Asians.

Enthusiasm for broadening the Asian–Pacific Economic Cooperating group (APEC) could be deepened when the US, Japan and China find meaningful ways to deal with the hard truths of country-of-origin labeling and off-shore sourcing, dumping goods at prices lower in the import market than in the home market, and theft of intellectual property (trademarks, copyrights and patents). Once China is a member of the World Trade Organization, some of these difficulties could be dealt with by the WTO. Right now they are the menu for bilateral trade relationships between the US and China with Japan, sometimes, acting as an unwilling mediator in these cross-Pacific disputes.

In the short run, at least for the next three–five years, Japan will probably back the US. After that, especially when the coastal Chinese economy (including Taiwan and Hong Kong) becomes more fully developed, Japan will probably defer to China. Both scenarios carry big ifs so it is a mistake to write in stone what will happen down the road.

Of course, free trade is an evolutionary concept for East Asia, too. Within Asean or South East Asia's customs union, free trade welcomes particular national cultures at different speeds, especially Thailand, Malaysia and Singapore. Also free trade plays a role in bringing stability to Indonesia, The Philippines and Vietnam. Moreover, free trade confirms failed nation-states, for example, Burma, Laos and Cambodia. Even with two decades of experience in free trade many of these South East Asian countries still restrict a long list of products from neighboring countries, and most have poorer regions (such as southern Thailand and northern peninsular Malaysia) that have not benefited from free trade.

Nevertheless, free trade is an idea that is spreading from Japan and the

Asean countries to China. Free trade transforms itself each and every time it is transmitted – from the lack of tariffs and the use of nontariff barriers in Japan, to the valiant attempts to take products off Asean's restricted list, to Korea's desire to reduced barriers in its services industry, and to China's commitment to export everything it can. Free trade within the Asian–Pacific region is now so contagious that India from South Asia and the Russian Far East also seek participation within an expanded APEC.

Free trade in slow motion

Very few leaders within East Asia start with the deductive knowing–believing paradigm. The American idea that free trade has universal application never was a crucial part of Japan's plan-rationale approach towards Japan, Inc., MOF, MITI, and overseas trade and investment. Japan's view about free trade was confirmed by Mexico, the failed nation-state within NAFTA. Yes the border industry program proved beneficial to Japanese investors who could re-export to the US. However, they had to live with the presence of rampant lawlessness among drug lords, the military and police, and government bureaucrats. And these Japanese investors had to live with the absence of control over local officials by the national government in Mexico City. Under NAFTA the north of Mexico has become a separate country unattached to the central institutions in Mexico City. Japanese investors wanted one national market, and they were unprepared for three regional markets – the Texas–Mexican north, Mexico City and the Indian south. In this case, both Japanese and American investors did a poor job in forecasting country risk in Mexico.

Japan's view about free trade was reinforced by eastern Germany's inability to close the productivity gap with western Germany, and the clear inability of the former Soviet bloc countries to move more quickly up the ladder towards long-term, sustained economic growth. The Japanese, Koreans and other East Asians know free trade thrives within a specific cultural and historical context, especially Anglo-American dominance of the international economy. Some East Asians believe the market-rationale approach towards the economy can be extended, in part, to continental Europe of the European Union and to a few export 'tigers' just off shore from continental Asia. Most of the world agrees with the Japanese, Chinese and other east Asians that the plan-rationale approach towards the economy can move their countries into contenders within the international economy.

Let's agree on the following: the specific cultural and historical context shapes the argument between the market-rationale and the plan-rationale approach towards the international economy. Therefore, in any discussion of East Asia, we must adduce answers to specific questions about society, culture, history, economics and leadership before we have a satisfactory explanation of how the logic of free trade shapes the nation-states of East Asia.

Japan, China and America

'Do we practice Confucianism?' The new president of Toshiba, Taizo Nishimuro, says 'Yes'. Here are a few of the management truths that he derived from his study of Confucius:[27]

- Recognize your own mistakes and have the courage to correct them.
- Expect your colleagues and subordinates to correct their mistakes.
- Keep your memory of the past in the present.
- Practice endurance to achieve goals.
- Evaluate core strengths and appreciate the strengths of others.
- Seek continuous self-improvement by ongoing reviews.
- Formulate ideas, win agreement among management and speed up implementation.
- Bring new ideas, people, alliances and partners to the worldwide company.

Even today Japan is so different for American and European executives. It has its own ways of doing business, many of which are derived from the writings of Confucius, their practice in China for over 1000 years and Japan's past and present Sinicization. Here is the unhappy truth for both Americans and Europeans: China's cultural power over Japan remains the most powerful force in how Japan and its business interests react to western proposals for reforming the international economy.

Free trade is not an obsession. Rather, free trade is a means towards increased exports, additional direct investments at home, and more suitable alliances, partnerships and joint ventures with foreign-owned firms. Free trade hides behind the name of globalization. Neither the Japanese or others in East Asia want a continuation of American hegemony over economic choices and business decisions. Although the American consumer culture has permeated many places in Asia, it is not seen as threatening to Confucianism, Buddhism and Islam as this same global, largely American culture overtakes local cultures in Canada, Mexico and Europe. Few in Japan and China see the American business culture threatening the local people's sense of identity.

In East Asia at least, the American ideal of free trade is reversible. Perhaps, the next few years are the make or break period for Japan's recent commitments to the US and its ancient ties to China. If forced to choose, Japan might defer to China.

Japan's big assignment

Japan's hesitancy to defer to China is likely go on for as long as possible. The collective memory of post-World War II Japan is more American-style freedom for individuals coupled with a government–business partnership to rebuild and reposition Japan within the global, largely American economy. Then came the bubble economy, and Japan, Inc.'s inability to protect the Japanese from asset deflation and many years of no economic growth. Today, Japan, Inc. is in the process of changing itself to make it more in tune with the demands from the US for more transparency and openness, but Japan could stop, say no more and look to China for its future.

Japan is not America, and Japan is not China. Japan is a different place, with a different history and a different people, culture and history. Japan's icons of memory do look different than those found in the US and Europe. Japan's icons of memory may look similar to those in China. This is an illusion. Japan's real cultural differences are there in the faces of its people who must survive on a group of islands with no oil, few minerals and not enough land for food.

Costs and benefits

How should Japan weigh the costs of attempting to deepen its commitment to the global, largely American economy against the possible, longer-term benefits of closer socio-economic ties to China?

This is a case of decision-making under uncertainty. If Japan continues with its political and economic commitments to the US, the possibility of additional difficult disputes between the two most important industrial powers within Asia and the Pacific would be almost nonexistent. The number is not zero, but it is low. If Japan seeks to staple its commitments to the US with some additional ties to China, the possibility of more disputes with the US and some with China would remain. How many more is unknown. If Japan defers to China, then all bets are off as to how the US will respond to this shift in power among the three most important countries in the region. In this case, the calculus for decision-making will have to be refigured to take in this new information.

Decision theory cannot create certainty where none exists, but it offers a way to organize one's thoughts and gather the information needed to make better decisions. Consider reimaging Japan, Inc. The best policy for Japan is to use these next few years to sort out where it wants to go and what it wants to become in the twenty-first century.

Remember Japan has choices. One option is to do nothing. Here is the interplay between Confucianism and decision-making under uncertainty:

- 'Keep the memory of the past in the present.' Do nothing or make small changes, and things will work themselves out in the end. This mastery of drift fits well into Confucianism; however, it finds few takers in the US. Japan may make this choice and find itself with an unhappy US partner. This is the downside risk.

Option two is to do something by making bigger changes. Here again are the links between Confucianism and decision-making under uncertainty:

- 'Recognize your own mistakes and have the courage to correct them.' Japan reforms itself from within. Japan, Inc., MOF and MITI, and *keiretsu* adapt themselves to changes in the international economy. Japan becomes a stronger competitor abroad in the US, Europe, China, South East Asia and Latin America.
- 'Expect your colleagues and subordinates to correct their mistakes.' Japanese business executives might find the strength to improve employee productivity, practice more efficient business operations and grow the national economy. Japan shows Korea, Taiwan, Singapore and others how to move away from an export orientation and build larger domestic markets.

Option three is to push for something so different that both the US and China find merit in the proposal. Here are the ties between Confucianism and decision-making under uncertainty:

- 'Practice endurance to achieve goals.' Japan brings China and the US closer together and knits the three countries into a new century of prosperity for Asia and the Pacific.
- 'Bring new ideas, people, alliances and partnerships to the worldwide effort.' Japan works with the US to create an Asian-Pacific free trade area. Both seek to extend NAFTA-type benefits to China, Korea and the Asean countries.

During the early 1990s, Japan pursued option one. However, Japan did not feel to good about it because asset deflation and zero economic growth were mugging the ability of its business firms to prosper in the demanding new world economy. Globalization overran Japan's ability to go slow in a world of faster decision making.

Then in the mid-1990s, Japan began seriously to think about option two. Each year Japan makes a few small changes to the way in which its government bureaucracy and business management reacts to full-scale changes in the US, China, South East Asia, Europe and Latin America. These Japanese initiatives are not enough. Too many managerial personnel remain on staff. Productivity suffers among Japanese business firms. The latter lag behind their American competitors in the race for competitive advantage abroad.

Today is a new day for Japan. Option three has one long-term political benefit. It puts Japan in the center of things considered important by the US – free trade with East Asia – and of things considered important to China – unimpeded access to US markets. Japan has a great deal riding on how well it makes its choices about the future.

Decision theory offers us a way to reflect on our choices and make the best decisions we can with the information we have in front of us. Decision theory offers Japan a way to organize itself to beat the fate of Salmon Day.

Salmon Day

Salmon Day is the experience of spending an entire day swimming upstream only to get screwed in the end. Softbank, Kyocera and Sony fought the global fight, gained market share and won support in Tokyo and on Wall Street. None of these firms were victims of Salmon Day, and they represent Japanese successes in the late 1990s.

Here are some summary thoughts. First, the transmitters of Japan's common knowledge about the big idea of co-operation with the US and China offer Japan the following: participation in the global, largely American life style with the chance that some of it might become a Japanese life style as well; English, Chinese and Japanese as languages of communication with the first having the primacy of use in the short run and the second having the primary of use in the long run; and Japan's ongoing big assignment – namely, the creation of a free trade within the Asian–Pacific region.

Second, the editors of Japan's common knowledge are in the process of reimaging their intellectual thoughts, conclusions and policy prescriptions concerning the U-World and the world of Asian loyalist nation-states. Many of their vivid memories of Confucianism will remain strong while others will be remembered as problems of transition to a more open Japan, Inc. and a wider opening among Japanese international firms. The costs and benefits of doing a better job of additional globalization among Japanese businesses depend, in part, on the quality of top management.

Third, managers and others in the professional class must decide whether Japan expands free trade and how it joins itself with the US and China in completing its big assignment in the twenty-first century. To decide for America means Japan continues to swim the North Pacific ocean without serious risk to its political and economic fate. However, to decide for America means Japan swims the ocean with a possible challenge from China. To decide for China means Japan faces a dangerous swim upstream and the possibility of getting screwed in the end.

All agree that Japan is essentially a political nationalizer. Japan uses nationalism to strengthen its global position within China and South East Asia. Although Toyota, Mitsuibishi and Toshiba are the beneficiaries of the Japanese plan-rationale approach towards the international economy, Softbank, Kyocera, Sony and Honda became successful overseas with pages opened in the market-rationale approach towards the global economy. Their Asian–Pacific strategies may be signposts to the future for Japanese firms.

Wisdom: decision theory lays out possible choices. Pay attention: one choice leads to Salmon Day, and the others lead to Japan sharing in a prosperous pan-Pacific twenty-first century. Open the doors: Japan makes a choice to defend its most important cultural interest without sacrificing its political and economic stability.

What do I get?

Here is our to do list:

- Buy into Japan's big assignment for the Asia-Pacific region.
- Accept loyalist nation-states as crucial to East Asia.
- Build on the memories of Confucianism, Buddhism and Shintoism.

- Reimage ideas about Sinicization and globalization.
- Recognize rooted cosmopolitans as 21st century Japanese.
- Create a pan Asia-Pacific stream of business.
- Use decision theory to weigh costs and benefits of options.

Remember these words: Salmon Day is not inevitable for Japan, but Japan will have to work with both the US and China to avoid Salmon Day.

Chapter 8

China Realism

'The best guess is that between a quarter and a third of apparent foreign-capital inflows may be mainland, or "false-foreign-devil", money.[1]

Dominic Ziegler, *The Economist.*

'Asian companies believe that the relationships comes first, and that investments flow from them. Relationship building is the essence of strategy, not a by-product of it.[2]

The Economist.

Strategic memory

Managing global business is about judging the certainty of China's realpolitik, its balancing of market forces and socialist practices, and its forthcoming decisions concerning Japan and the US. Is there a crisis? Here is the risk assessment facing business executives.

China's strategic culture

According to Alastair Iain Johnston, a political scientist at Harvard University,[3] China cloaks its intentions within Confucian rationalizations, but ranks its strategic preferences according to several realpolitik assumptions about the nature of future conflict with an enemy. For example, China uses noncoercive, delaying negotiations when confronting a more powerful enemy, and then uses offensive, violent means when China is sure of annihilating an enemy. Debate among Chinese policy makers under the Emperor, nationalists and communists is only over whether China is strong enough, in military terms, to prevail. Both the Japanese and others in East Asia believe China will be expansionist during the twenty-first century.

China commonwealth

'Greater China' means different things. It refers to the 1.2 billion Chinese on the mainland and in Hong Kong and Macau, and another 21 million in Taiwan. Also it refers to 'the 50m-odd ethnic Chinese, mainly in South East Asia, whose wealth ... equals that of China.'[4] Moreover, it refers to cultural, trade and investment links across the Pacific Ocean among the international Chinese community in the US, Canada, Latin America and Australia. Note that China has economic problems. First, its 7–8% growth rate is over overstated by official statistics; second, its domestic economy is over-regulated; and third, China's national economy is divided between an upbeat coastal consumer society and a poverty-stricken rural community. Therefore, neither the Japanese or others in East Asia anticipate China becoming the dominant economy power in Asia during the twenty-first century.

Western China

Four-fifths of mainland China is in its western frontier lands. Oil and the bulk of China's minerals are in these rugged lands of Xinjiang that are home to non-Han Chinese people, the Moslem Turkic-speaking Kazakhs, Kyrgyz and Uighurs, and the Moslem Farsi-speaking Tajiks. These minority people want political ties to the newly independent Central Asian nation-states of the former Soviet Union. Even in Tibet and Inner Mongolia nationalism is on the rise among the native people. All told about 120 million minority people are contesting who benefits from how the Han Chinese govern China. However, no one believes these internal problems will spill out over the borders of China into Russia, India, Korea and Vietnam, or into the sea lanes vital to Japan, America and South East Asia.

China's investment strategy

In the eastern one-fifth of China live two-thirds of China's population, predominately rural and truly poor who now have rising expectations from the TV shows on life in the wealthier coastal cities of China and elsewhere in the China commonwealth. Although a great deal has been made of how foreign money (from Taiwan, Singapore and elsewhere in South East Asia, Japan, the US and Europe) has transformed Guangdong, Shenzhen, Fujian, Shanghai, Dalian and other cities, 'the best guess is that between a quarter and a third of apparent foreign-capital inflows may be mainland, or "false-foreign-devil", money ... Exports from foreign ventures ... is about 30%, and growing voraciously ... [but] a lot of that consists merely of importing components, bolting them together and shipping them out again ... [Worse yet] any gains in efficiency, management and technology that foreign ventures may be enjoying are not spilling over into the state-owned sector ... the 'commanding heights' of the Chinese economy.'[5]

China reform

'In his first major policy initiative since Deng Xiaoping's death, Chinese President Jiang Zemin is set to open the door to a reform Mr Deng didn't tackle: privatization. No-one in China is calling it that. But Mr Jiang and other leaders have reached a consensus to allow most state-run enterprises to diversify their ownership by taking on domestic institutional investors, issuing stock or allowing foreigners to acquire stakes.'[6] No longer will the state sector dominate the domestic industrial output and take the lion's share of China's resources. Instead, foreigners will be allowed to participate in energy, services, infrastructure and telecommunications. However, note the twist: the 1000 largest state enterprises will be saved and their performance will be monitored by outside 'human resource centers', but the mid-size and 240,000 small state firms will be allowed to sink or swim.[7] If all goes well, 'China is seeking to build industry groups which become powerful conglomerates in their own right, free to raise capital on domestic and international markets ... [According to Chinese prime minister Li Peng,] ... state enterprise reform is not only a ... major economic issue ... [but] also a major political issue of vital importance for the destiny of the socialist system.'[8]

Socialist memory

Socialism with market characteristics was Deng Xiaoping's idea. He asked these questions: is China 'fording a river by feeling for the stones' – that is, is China's plan-rationale approach of gradualism slowly integrating the mainland into the wider international economy? Or is China 'an inert system, rolling under its own weight' – that is, does China need a new impetus from the whole of the China commonwealth to transform itself into a competitive partner within the international economy? Or is China still a 'socialist system with market characteristics' – that is, are the reforms by the leadership sufficient to ready China to face the global, largely American world economy?

Deng forged answers to these questions because most of his 'peers were nostalgic for the "golden age" of the 1950s, the Soviet model redux but reshaped. Deng forced them to go in the opposite direction ... He was a demolition man. Deng deconstructed the China he took over, not the traditional China of Confucian values and Taoist cults, but the China of Communist principles and practices which he had himself helped Mao Zedong superimpose upon their land ... Deng adopted and adapted the East Asian [plan-rationale] model ..., unleashing forces rather than mobilizing them ...

'In retrospect, [Deng] seems a Janus figure, with one face looking forward, smiling upon the brilliant economic future which he opened up for China, but the other grimly set upon the past, looking back to authoritarian political control by a bureaucratic elite ... But Janus was also the god of doorways and of all beginnings, and the "paramount leader" will surely be viewed as the man who finally opened the right door for the journey to begin, even if he refused to go on to the end.'[9]

Governing the market

Governments with a strong view of governing the market admit foreign-owned technology when it is good, desirable, and low cost. For example, 'James Chu, a US-trained computer scientist, is helping China defend its sovereignty against ... barbaric information. [He] runs a small Hong Kong company called China Internet Corp. With backing from Xinhua News Agency [China's official source for news], Chu is pioneering what he calls the 'Intranet,' an insulated, but not isolated, corner of cyberspace ... Chinese users would have unlimited access to each other, but only screened links with the world beyond,'[10] says Chu.

Socialism with market characteristics also reverses ideas about capitalism. For example, both public and private economic information must work on behalf of society. Second, information management must promote values held dear by the nation-state. Third, business transactions that uphold the authority of the ruling party supersede all others; then come family transactions through extended networks with overseas Chinese; and finally come deals with Japanese and American investors. Fourth, obscurity and caution are inherently better than formal laws and predictability to encourage party and family insiders to do a better job for China and keep foreigners at bay. Socialist memory has become capitalist practice.

For the last 20 years, which coincide with the rise of Deng Xiaoping to paramount leader, China experienced the emergence of a private sector with strong ties overseas, and the growth of a middle class. This path towards economic development fits within the ruling party's desire to avoid politicizing the national economy, such as the experience of South Korea during the same period of time. Also this passage towards market expansion suggests a limited answer to the question of whether the national government today is capable of running something as sophisticated as Hong Kong and Macao as they return to China in 1997 and 1999, respectively. Moreover, this journey towards global economic integration in the twenty-first century points to a 'one country, two systems' covenant hinted at by mainland China for Taiwan.

China's ruling party created a social infrastructure in which the families of the party elite go in and out of the private sector. For example, they use stock listings on the Hang Seng Index (Hong Kong) or in Shanghai to promote business activities for the state's 'red chip' and privately owned firms. Also these leading families stay tied to the emerging middle class whose energies are driven by good ideas from local entrepreneurs, returning foreign-trained citizens, newly rich overseas Chinese, and Singapore's government and commercial elite.

Foreign investors also come to terms with the ruling party because the latter does not expropriate the former. And the government comes to terms with family-owned conglomerates and global firms because the former needs the technology, capital, and managerial capabilities of foreign investors. In the mid-1990s, 'Beijing is scaling back the preferential treatment that it extended to foreign investors … It recently launched a crackdown on foreign companies that are evading taxes, and withdrew foreign firm's rights to import a range of personal items – from photocopy machines to mattresses – duty free … The thinking is that access to China's markets, workforce and resources should be enough to keep drawing foreign funds.'[11]

Socialism with market characteristics means the ruling party in government plays a major role in the management of the national economy. For the foreseeable future, China's approach towards governing the market is that country's long-range social and economic trend. Chinese socialism is different from the habits of capitalism found in the US, Europe, Japan, the Asean countries, India, and the rest of the world.

China's big assignment

Notwithstanding these ongoing difficulties foreign firms make investments and build market share in China. Business executives collaborate with local state-owned enterprises, and with privately owned firms whose foreign alliances have been approved by the Chinese government. These co-operative relationships organize immobile local assets, make them more productive and, if successful, embed them within the industrial structure of the nation, region and city.[12] Foreign firms select those objectives that are appropriate for China. Here is a short list:[13]

- Conform to the practices of an industry by using equity or contractual joint ventures.
- Compare the costs of importing newer technologies versus using existing local assets.
- Accept the limitations of the infrastructure in the host country.
- Balance the role that political ties to the ruling party, their families (*taizi dang* or Princes of the Party), and national and regional authorities play in revitalizing existing plant and equipment, or approving the building of new factories.
- Embrace government–business and interfirm alliances.

These are recipes for successful foreign investments in China.

China as a loyalist nation-state

The orders of the Chinese Communist Party and the subsequent investments by the Chinese state-owned firms present hard evidence about how China governs the market. The Chinese press call state-owned trading and industrial conglomerates 'aircraft carriers'. Often they are 'headed by former government officials, have smooth access to bank financing, stock-market listings, and regulated areas of industry and trade'.[14] These attributes make them hard to turn around even when they lose money manufacturing goods customers don't want. Of course, profit and loss is not the top priority among Chinese 'red chips'.[15] This attitude affects the success of alliances with foreign-owned firms.

For example, the stock of Tsingtao (pronounced chingdow) fell by 57% in 1995 because the $190 million (in US dollars) raised by the brewery was not invested in expansion, but lent to other unidentified Chinese companies, state companies immune to repaying their debts.[16] Anheuser–Busch pulled out from raising its stake in Tsingtao to 15% and instead 'went ahead with a joint venture to produce Budweiser with a different brewery in the central Chinese city of Wuhan'.[17]

Through 1995 foreign financial 'investments flow in, but the returns do not flow out ... [because investments are] being redistributed to other elements in the economy, which are obviously better off',[18] says William Kaye, managing partner of Asian Hedge Fund in Hong Kong. The World Bank suggests what has happened to these funds: 'A dearth of [public funds], immature legal, banking and social-security systems and, above all, fear of social chaos are blocking the way to reversing debilitating losses in China's huge state-owned industries.'[19] Clearly, between 1992 and 1995, 'Chinese equities available to foreigners have proved to be one of the world's least profitable [financial] investments, losing 60% of their value'.[20] Foreign capital supports the old order – that is, state-owned firms run by family members from the ruling party.

How do local privately owned firms prosper in the Chinese market? 'The most successful [Chinese] entrepreneurs share a common strategy: They prey on less-nimble state companies [by acquiring the latter] ... and turning them into private conglomerates ... While the state sector sputtered along with little real growth, companies without a state ownership expanded their output at a whirlwind rate of 24%.'[21]

Private entrepreneurs rely on loans from family and friends to get started, and then expand only with retained profits, always struggling on their own,[22] says Huan Guocang, an economist for J. P. Morgan & Co. For example, in 1982 the Liu brothers of Chengdu in China's Sichuan province, '"jumped into the sea", as the Chinese call shifting from the public to the private sector, raising US$500 by pawning their bicycles and watches. They started raising pigs, quail, chickens and ducks ..., [now] own 40 animal feed mills scattered around 20 provinces in China ... [and] are one of China's first centimillionaire families.'[23] Today, the Liu brothers compete against Thailand's world-class Charoen Pokphand Group (CP) and the inefficient government-owned feed mills.

For the moment, growth in China's foreign trade is dependent on the state sector's ability to make prosperous alliances with foreign-owned firms. Of course, China's future depends heavily on its entrepreneurs doing deals with overseas Chinese and global firms.

China and the U-World

China competes against the South East Asian countries and India for additional foreign direct investments. As China seeks integration with the international economy through the World Trade Organization (WTO), it wonders whether it will continue to enjoy an increase in overall real income.

Protectionism

Right now the US, Japan and Europe prohibit Chinese worker mobility, maintain high tariffs and nontariff barriers on many Chinese exports, and they limit inward investments to those projects whose high returns are greater than the accompanying high risks.

Sovereignty

Also China itself limits the flow of foreign capital to approved joint venture projects with 'red-chip' or privately owned enterprises, and it inhibits the flow of goods into export markets by its sloppy protection of foreign-owned intellectual property. These are the long-term conflicts between China and the countries of the industrialized world.

If these nonoptimal barriers to trade are removed will China be better off at the expense of the US, Japan and South East Asia? That is, will the industrial countries suffer at the expense of the less-developed countries, or will China suffer at the expense of the wealthier nation-states? Also will northern, middle and southern China all be better off, or will one region be a winner and the others losers?

The industrial countries have the advantage of scale economies over China. They can produce in volume, and keep margins up over the product-development life cycle of their goods. So long as they can ship into Hong Kong (for southern China), Shanghai (for middle China), and other coastal ports, their transport costs are low, too. Once the industrial countries (and their global firms) seek to move goods inland within China, they do not have the same advantages because rail, truck and air freight – all managed by state-owned firms – are unreliable, destructive and expensive.

Nevertheless, southern and coastal China are making the same transition from agricultural suppliers to manufacturing exporters as did South East Asia. Southern and coastal China, which were peripheral to and had exogenous roles within the world economy, are now taking on endogenous roles – namely, building up internal market demand for local and foreign goods. Within southern China, transport costs are falling as new expressways are built on both sides of the Pearl River, new airports are constructed in Macao, and new ports are built in Hong Kong. The same is going on in middle China (in and around Shanghai) and in Manchuria. But the bottlenecks remain in moving goods at the lowest cost possible among the many regions of China.

Therefore, southern and coastal China will continue to create external economies that may lead to growth of regional industrial activity at the expense of peripheral regions in northern and inland China. The former's advantage will grow based upon the input–output linkages among 'red chips', privately owned firms, overseas Chinese conglomerates and global enterprises. These firms are responsible for both backward and forward linkages, and the introduction of new technology for the long-term improvement of the region's competitiveness within China and in the international economy. Also the region's advantage will grow as local transport costs fall and even more industry concentrates itself in the south and along the coast. Moreover, the region will continue to have low wage costs because Chinese from inland will crowd southern and coastal China looking for nonpeasant, urban work.

China offers two incentives for moving manufacturing from the core countries to its southern and coastal regions:

- *Scale economies.* The opportunity to be in on the ground floor of building up production for a potentially vast internal market.
- *Transport costs.* The end of remoteness from world markets, suppliers, and customers, and perhaps from internal markets, too.

Given labor immobility between China and the industrial countries China's wage rates will remain low. The only way to slow down China's rush towards industrialization is for the industrial countries to keep their tariffs high. Once China is in the WTO, the US, Japan, Europe and South East Asia will lose the tariff weapon. Then China will be a fierce competitor for domestic and export market share in the twenty-first century.

Free trade in very slow motion

Successful foreign investors handle these China problems by measuring risk against the contextual forces of five-year plans, infighting among party leaders, forced investments by 'red chips' in listed Hong Kong companies, and family ties across the China commonwealth. Macao-based *compradores* (or import agents), Hong Kong-based trading houses (so-called Noble Houses), family owned conglomerates of overseas Chinese, and a few global firms are masters at developing a cultural construction of risk associated with doing business within mainland China.

On the other hand, less successful foreign investors examine the risk, set benchmarks for long-term performance, and conclude that it is safe to go ahead. As problems mount, as returns deviate further from financial goals, this group of foreign investors convinces itself that it now has a better understanding of 'the Chinese market'. The difficulties are explained away, or they are found not important enough to interrupt the friendly relations between foreign investors and government.

Thus unsuccessful foreign investors slip imperceptibly into accepting normalized exceptions as cultural norms for 'the China market'. They tend to be arrogant about what they don't know about the politics of China. Western corporate culture tends to breed hubris among individuals even when their failure to question decisions of Chinese organizations is exposed. Their inward emotion creates a relentless inevitability for mistakes to happen again and again. They think they can beat the odds sometime in the future, but this is only false confidence among unsuccessful foreign investors. Instead, they lose even more money in China.

Difficulties are inherent in all investments in 'the China market.' The potential for failure is always possible. The cultural risks are not easily manageable, the policy changes of the ruling party are not easily preventable, and the rituals of financial disaster for privately owned firms have no meaning within socialism with market characteristics. Some combinations of minor failures might lead to catastrophe for even the very successful trading houses; right now a few are acting on Hong Kong's return to China by delisting their stock on the Heng Seng and moving their corporate headquarters to Singapore. Less successful global firms show bafflement about why things are not going as planned in China, and they compensate for lower risk in southern China by taking greater risks elsewhere in China. What failures like the 'Beijing Jeep' should teach us is that we have constructed an investment world in which the potential for losses is embedded in the fabric of day-to-day life for foreign investors in China.

Successful global firms use the know-how they acquire at home and in Japan, South East Asia, Taiwan and Hong Kong, and apply it to their expansion in China. Sometimes, their initial entries are in 'across the fence' factories in the special economic zone of Shenzhen in which goods are assembled for export to the US. Then they may ship goods internally within southern China. If they are successful, they might then set up shop in Shanghai or Manchuria, or in both places. These global firms recombine their knowledge as best they can because business conditions are different in Cantonese-speaking southern China versus Mandarin-speaking middle and northern China. The former tend to have family ties with overseas Chinese throughout South East Asia and in the US; the latter depend heavily on nonfamily alliances with investors from Singapore, Japan and Korea.

No doubt American-owned global firms possess superior knowledge about American policies on foreign trade, the willingness of the US to transfer technology, and opportunities for exports of Chinese goods to the NAFTA countries. If these global firms combine with overseas and southern Chinese family-owned firms, American firms in China can cope better with future asymmetric shocks from the ruling party and other Chinese organizations. Without such alliances, American-owned global firms are bound to fail because they possess little good knowledge about the inner workings of China.

What are the results in human terms? Since the early 1970s, 'newfound plenty spread across the land: grain consumption grew by 20%, intake of seafood and pork doubled, Chinese ate three times as many eggs and quadruple the amount of poultry. And in urban areas, an emerging consumer society was quickly established. In the early 1980s a bicycle, a wristwatch or a transistor radio was a consumer status symbol. Fewer than one urban household in a hundred owned a color television or refrigerator; only six in one hundred had a washing machine. A decade later, there are more than 70 color television sets, 80 washing machines and 50 refrigerators per hundred urban families ... [This] concrete improvement in living standards makes ... the figure of 9% annual economic growth for 14 years ... impressive ... [and] more real,'[24] reports the Guinness Flight China & Hong Kong Fund.

Productivity and performance

During the 1980s and 1990s, 'economic growth in China ... has been outstanding by international standards. The average yearly growth for gross domestic product ... was about 10%, according to the World Bank. [China has had] technological progress, and changes in the efficiency with which existing technology is applied to production, i.e. technical efficiency ... Technological progress dominates technical efficiency changes as the main source of total factor productivity growth in the [state sector, rural industry, and agriculture].'[25]

What does this mean? More production occurs in China today, but the wide gap between standard Chinese practice and best international practice remains throughout most of the country.[26] Only the coastal region begins to close the gap in technical efficiency between China and the world in the rural industry and agriculture, but not in the state sector. Unfortunately, 'the contribution of technical efficiency change to total factor productivity in all regions is minor, as it is in the sectors'.[27] This means that China is *not* catching up with the South East Asian countries, Korea and Japan, even though global firms from the US, Japan and Europe are increasing their investments in China.

'Despite the impressive expansion of exports out of China, the large inflows of foreign direct investment into China, and the increasing access to international capital markets, it is still premature to judge that China's foreign-sector reforms have been a stunning success. China's rapid export growth has depended to an unprecedented degree on foreign-invested firms. [When this reliance on foreign firms is] combined with the protection provided to state-owned industries, [such reliance] has inhibited productivity growth, especially in intermediate-input industries where prices are still above international levels. Thus rapid export growth from foreign-invested firms, a large share of which is export processing, has only limited links back to the domestic economy, and the domestic content of exports is low. To some extent, export industries appear to be enclaves, and China's state-owned industries lag far behind in export growth. In the absence of widespread industrial restructuring, China will be hard pressed to sustain the growth of exports at a rate anywhere near that of the past decade.[28]

Win–win–lose world

Do global firms make money in China? 'The Economist Intelligence Unit and Andersen Consulting found profitable foreign companies in every business sector, from consumer products to telecommunications ... Bank of East Asia researchers report a similar trend ... [According to the Andersen report], three-fifths of foreign-funded ventures operate in the black, and they tend to become profitable in less than two years. Similarly [in the Bank of East Asia report], half ... [made] profits there and 85% said they were pleased with their investments.'[29] Today, foreign investors earn back their start up costs in under five years, a record time for China.

The U-World is at the end of the beginning of a new win–win–lose world market era for foreign investors in China. Here are the crucial questions for foreign CEOs:

'What do we get?' New joint ventures and alliances with well-connected family- and state-owned firms. New knowledge, better products and lower costs. Application of cultural assets to real business problems. Foreign cor-

porate leaders have opened up Chinese factories in the race to build market share within China and export China-origin goods overseas.

'What can we do?' Be a prophet about the China market. Forecast correctly about long-term changes in southern and coastal China versus the vast hinterland of China. Become a part of the networks of suppliers to other global firms. The CEOs are taking their firms on a voyage of unprecedented difficulty as they move away from Hong Kong, Shanghai and Manchuria into the interior of China.

'Tell me what you are going to do on Monday that's different?' Let's look at several examples.

Motorola

With $1.2 billion (in US dollars) invested in China, Motorola, a US high-tech giant, is that country's largest investor. Its newest plant is the $560 million semiconductor fabrication plant in Tianjin. Motorola is betting that by the year 2004 'China will emerge as the world's largest consumer of electronics products – everything from PCs and televisions to microwave ovens and telephones ... [and] by the year 2005 or 2010, China will have the world's largest electronics industry'[30] says C.D. Tam, head of Motorola's semiconductor products group in Asia.

These are the key 'What can we do?' questions facing Motorola executives in China:[31]

- Inculcate Chinese employees with Motorola's corporate culture – an unwavering obsession with quality – that is known throughout the world as 'Six Sigma,' or fewer than four defects per million units.
- Popularize the cellular phone, builds brand equity, and turns itself into market leader.
- Manufacture pagers and cell phones, and packs semiconductors into housings for both domestic and export markets.
- Go into a joint venture with a state-owned enterprise, Nanjing Panda Electronics, to produce a personal computer based on Motorola's Power PC chip, the major rival to Intel's Pentium.

- Supply components for cars assembled by the alliance between General Motors and Shanghai Automotive Industry, one of China's biggest car makers.

And these are the 'What do we get?' questions for Motorola in China:[32]

- Sales doubled over two years to reach $3.2 billion (in US dollars) in 1995 – almost 12% of Motorola's worldwide revenues.
- Three million new cellular phones sold every year until the year 2000.
- Income comes from the sale of chips, pagers, and cellphones.
- Money is lost on PC sales because Chinese distributors fail to pay their bills.

What's ahead for Motorola in China? Today, it is the dominant American investor, twice as large as Atlantic Richfield ($625 million invested), ten times as large as Lucent Technologies ($150 million invested), and twelve times as large as Hewlett-Packard and IBM (each with $100 million invested).[33] Tomorrow, China will be one of the fastest-growing markets in the world. That's a good bet for Motorola.

Lucent

Although the investment of Lucent Technologies (formerly, AT&T Network Systems) is one-tenth the size of Motorola's investment in China, Lucent has put together a different package of benefits for its Chinese partners. The firm co-operates with China in ten main areas – namely, digital switches, microelectronics, network management, optical transmission products, training systems integration, and research involving Bell Laboratories.[34]

As early as 1991 Lucent had the following interests in China:[35]

1. The firm is a main investor in the $12.6 million (in US dollars) Beijing Fibre Optic Cable Co., a joint venture to produce fibre optic cable in China.

2. The firm is a partner in the $9.82 joint venture with Shanghai Telecommunications Equipment Factory to produce digital subscriber-loop carrier systems.
3. The firm sells $7.5 million in mobile telecom equipment and switching equipment to remote Chinese provinces.
4. In 1993 the firm began building high-tech phone switches and it made plans to help China make cellular phone equipment.

In the mid-1990's, the firm announces plans to build China's best-organized and equipped, fully automated optic-fibre manufacturing plant in Shanghai, and to upgrade the manufacturing operation for switching systems in Qingdao.[36] In the latter partnership, the firm will transfer to Huajing Electronics Corporation Group, an affiliate of China's Ministry of Electronics Industry, technologies for processing and designing very large-scale integration devices. With these deals Lucent's total foreign direct investments in China totals $150 million (in US dollars).

Microsoft

> *'Over the next five years, China plans to invest billions to computerize its immense, paper bound financial sector, government bureaucracy, and state industries. As the Chinese also rush to buy personal computers for their homes and small businesses, PC sales are expected to explode from 1 million units in 1995 to 5 million in 2000, according to International Data Corp. Microsoft is clearly positioning itself to be king of China's software industry.'*[37]

Here are the key 'What can we do?' questions facing Microsoft executives in China:[38]

- Put in place a massive-on-the-ground consulting presence.
- Train technicians at research centers and universities.
- Develop a Mandarin-language version of Windows designed especially for China's simplified characters.
- Put PCs using Windows and DOS into government ministries.

- Work with Chinese researchers to develop interactive TV, and speech and handwriting recognition programs.

And these are the crucial 'What do we get?' questions for Microsoft in China:[39]

- Share design codes with institutes connected to China's Electronics Ministry, which also supervises state-owned software companies trying to develop competing operating systems.
- Establish Windows as the national platform for China.

What's ahead for Microsoft in China? More piracy of its software, and more local (still weak) enforcement of China's intellectual property laws. Available in Chinese cities 'is the mother of all pirated software packages – a CD-ROM known colloquially as *Da ... bu tie*, or the "the almighty medicine," which incorporates no fewer than 80 different software applications, including Lotus, Microsoft, and Novell products. These software packages would cost $8000 (in US dollars) in the West. In China, they are sold for less than $100.'[40]

Do the foreign direct investments of Motorola, Lucent, Microsoft, Alcatel and others boost manufactured exports from China to Korea, Taiwan, and the South East Asian countries? In the mid-1990s, this is still a goal rather than reality. However, these investments have had positive employment effects, and they have augmented China's smaller domestic capital base. Moreover, China has made every effort to exploit foreign sources of technology, and these will increase as Chinese state-owned enterprises do more joint ventures to maintain China's access to foreign markets.

Inward foreign direct investment remains the primary source of new funds for building the domestic internal market in China. The country is different to its neighbors in South East Asia. They grew and prospered through export-led development. On the other hand, China is developing its economy by building up its internal market with the help of foreign capital and technology, a great deal from networks of ethnic Chinese whose families own many emerging global enterprises from Thailand, Malaysia, Singapore and Indonesia, and smaller international trading firms from the US, Canada and Australia.

Network citizenship

Here we have China realism in a nutshell. East 'Asian companies believe that relationships come first, and that investment opportunities flow from them'.[41] According to Tsun-yan Hsieh of McKinsey, a management consulting firm, 'Relationship building is the essence of strategy, not a by-product of it'.[42]

What are its deep roots? Many East Asian 'firms were established by migrant Chinese or their offspring; they build up networks in which extended families and clans did business only with one another in order to reduce risk. The interlocking relationships soon began to extend to local officials ... This web of connections, or *guanxi*, still sits at the heart of most overseas-Chinese groups.'[43]

Networking in Asia is a form of protection for American and other Western firms. In most East Asian countries, relationships are the only form of commercial security. Most rules and regulations remain to be written by local governments; those that are written are not always enforced by local bureaucrats; and even if they are written and enforced, most markets in Asia and China simply lack the transparency of Western markets. Thus Western business executives do not always know what is going on, who are the crucial decision makers, and how government and business will share the spoils of new investments.

Is there a way out? Form partnerships with local governments and businesses. China demands joint ventures for many foreign investments in strategic sectors of the economy. Korea, Malaysia and Indonesia strongly suggest foreign alliances with local partners. And Thailand and Japan offer global firms the opportunity to join with local government and business partners to enhance the success of foreign investments.

Networks of relationships among families, friends and government officials always win. Foreign firms must cultivate the necessary connections before deciding which businesses or projects to start up in China.

China redux

'The opening of the China market and its internationalisation means the biggest challenge for Chinese companies is competition from overseas and from multinational companies involved in joint ventures',[44] says Duan Yongji of Stone Group. The firm is collectively owned, and it is the leading Chinese producer of electronic cash registers, important in linking up banks, credit agencies, and retail outlets. Stone Group is repositioning itself to team up with global firms to sell personal computers (Compaq), integrated circuit chips (Mitsubishi Electric and Mitsui), and to be a player in 23 other joint ventures. Clearly, Stone Group and other Chinese companies obtain their new technology through joint ventures with foreign-owned global firms. The import of advanced technology has been driving China's economic growth, but output per unit of input is still not rising. That is China's great worry.

Even in foreign-owned retailing productivity and performance have stalled. For example, Yaohan, a Japanese retailer, planned to build 1000 supermarkets and department stores within China. As a consequence, it moved its group headquarters to Shanghai. Unfortunately, 'the Chinese trooping through its smart new store are only window shopping. Average incomes are low … Just 120m or so urban Chinese are rich enough (with an annual income of more than $1000) to afford even such modest items as detergent or packaged food … [If they buy] luxury goods, household products or clothes, the Chinese are likely to choose a style or brand they judge [conforms to their] desire to conform.'[45]

Thus China is not jumping ahead of its neighbors, Korea, Taiwan and the South East Asian countries. China's miracle is still more myth than reality.

Costs and benefits

How should China weigh the costs of attempting to deepen its commitment to the global, largely American economy against its desire to remain a socialist society with market characteristics?

This is a case of decision-making under uncertainty. If China continues with its gradual reform of state-owned firms, the possibility of further rapid rises in standards of living, of increases in productivity and of gains in economic growth is low. If China accelerates reforms by selling off more state assets, collecting delinquent taxes and allocating funds for improvements in energy, distribution, transportation and telecommunications, the possibility of strong gains is higher. However, if reform goes too far too fast, socialist memory might come back to oust the post-Deng leadership. In this case, the calculus for decision-making will have to be refigured to take in this new information.

Decision theory cannot create certainty where none exists, but it offers a way to organize one's thoughts and gather the information needed to make better decisions. Consider reimaging China. The best policy for China is to use these next few years to sort out whether it wants to continue with gradualism or accelerate reforms of state institutions.

Remember China has options. One option is to go slow and continue with the policy of gradualism. Most Communist Party members and the military will go along, because this will 'Keep the memory of the past in the present'. However, this policy will not satisfy those in the China commonwealth who want faster change to accommodate Hong Kong and to entice Taiwan back into China.

A second option is to accelerate reforms. It seems the post-Deng leadership under President Jiang Zemin has decided to choose this option and 'Recognize their own mistakes and gather the courage to correct them'. Most Communist Party members will maintain a wait and see attitude towards these reforms. The military, especially those in the People's Liberation Army who gain from managing state assets, and leaders in the China commonwealth will work to make this policy a success.

A third option is very rapid reform. This is a push for something so different that it amounts to a wholesale dismantling of the socialist system. Today's China has the example of the break-up of the Soviet Union and the dismantling of the Russian Communist Party's control over the economy. China does not want its minority population to become independent, and the Chinese Communist Party does not want to give up control over the market and government.

All post-Deng leaders will be judged by how well they reform China's domestic economy within the boundaries set by the Communist Party.

Today, China is pursuing option two with due regard for the sensitivities of those in the Communist Party who are skeptical about an accelerated reform program. Will China be able to do this successfully? Yes. None of the members of the post-Deng leadership or the new industrial elite want to deconstruct Deng Xiaoping's China. They fear the chaos of violent revolution from the rural and urban poor, an outcome that would oust the Communist Party from its leadership position.

Decision theory offers us a way to reflect on our choices and make the best decisions we can with the information we have in front of us. Decision theory offers China a way to organize itself to beat the fate of Salmon Day.

Salmon Day

Salmon Day is the experience of spending an entire day swimming upstream only to get screwed in the end. Motorola, Lucent and Microsoft are fighting for market share in China. They are doing what 'old China hands' from the US did unsuccessfully in the 1980s, and what Japanese investors, on the whole, have done well in the last two decades. China is one of the emerging markets of the world in which the victims of Salmon Day outnumber those who were successful in establishing a position of strength in the China market.

Here are some summary thoughts. First, the transmitters of China's common knowledge about improving the socialist system offer the following: participation in the East Asian, largely Chinese life style with the probability that some of it might become an American life style as well; Mandarin Chinese and English as languages of communication, with the first having the primacy of use in both the short and long run; and China's big assignment – namely, accelerated reforms of China's domestic economy.

Second, the editors of China's common knowledge are in the process of reimaging their intellectual thoughts, conclusions and policy prescriptions concerning the U-World and the world of Asian loyalist nation-states. Confucianism and communism remain the two strongest intellectual currents whereas Western and Japanese ideas continue to be weaker ideas among those used to underpin the communist leadership's commanding role in society. The costs and benefits of doing a better job with the domestic economy depend, in part, on the quality of foreign partners from South East Asia, Japan and the US.

Third, managers and others in the professional class must decide whether China invests in itself or spends profits on luxury and leisure goods. To decide for investments means China improves the productivity and performance of its enterprises, grows the economy and raises up the standard of living of the urban middle class and the rural poor. To decide for spendthrift consumption means China stays as it is with little opportunity for raising up the rural poor, broadening the urban middle class, and translating the success of Hong Kong and Taiwan to mainland China. To decide for consumption now over longer-term investments means China faces a dangerous swim upstream and the possibility of getting screwed in the end.

All agree that China is a political nationalizer. China uses dual, identity and network citizenship to strengthen family connections and personal relationships among the peoples within the China commonwealth. Although Motorola, Lucent and Microsoft are the current beneficiaries of the Chinese plan-rationale approach towards the international economy, no-one among the 'old China hands' will be surprised when they fail to alter the way in which business is done in China. Unfortunately for investors, both American and European executives tend to have a forgetfulness about the great risks and low returns from past and present investments in China, and they have a mental block about the real political and cultural reasons behind why the Japanese and South East Asians make similar investments in China. Japan's China strategies are not signposts to the future for American firms.

Wisdom: decision theory lays out possible choices. Pay attention: one choice leads to Salmon Day, and the others lead to China sharing a leadership role in the economy of East Asia. Open the doors: China makes a choice to reform itself without sacrificing its political stability.

What do I get?

Here is our to do list:

- Buy into China's big assignment for domestic reform.
- Accept China as the most important loyalist nation-state within East Asia.

- Build on the memories of Confucianism and the socialist system.
- Reimage ideas about socialism, capitalism and globalization.
- Recognize middle-class city dwellers as twenty-first century Chinese.
- Create a pan-China commonwealth stream of business.
- Use decision theory to weigh costs and benefits of options.

Remember these words: Salmon Day is not inevitable for China, but the post-Deng leadership will have to work very hard at reform to avoid Salmon Day.

Chapter 9

Paradigms for South East Asia

'Intense dependency on the extended family [is] a social reality, which is vital to the Chinese, and not just a desirable goal.[1]'

Andrew Adonis, *Financial Times.*

'In Asia we agree to things only if they are subject to change without notice ... If we want the governments to endorse the plan, then it is important to make it clear from the beginning that they will not have to follow it.[2]'

Narongchai Akrasanee, chairman of General Finance & Securities, a brokerage company in Bangkok, Thailand.

Citizenship memory

Dual, identity and network citizenship creates business success in East Asia.

First, networks. Local executives and entrepreneurs grow long-term sales, profits and returns on investment based on family connections, and networks of commercial friends and government contacts. Chinese-owned firms from the China commonwealth and South East Asia strengthen family and personal relationships before they take advantage of new investment opportunities offered by global firms from Japan, the US and Europe.

Second, passports. Local business people fashion short-term triumphs based on political commitments to how individual nation-states govern the market. Chinese-owned firms throughout East Asia use their web of connections, or *guanxi*, to protect themselves and their foreign partners from shifts in local government policy about savings versus consumption, investment targets, and repatriated profits versus national taxes.

Networks of relationships among families, friends and government officials always win. They are crucial to understanding the ins and outs of the plan-rationale approach to the economy. Successful foreign firms cultivate the necessary connections before deciding which businesses or projects to start up in China. Talented foreign business executives ask the right questions:[3] 'What do *they* want to do? What are their objectives? Their values? Their ways of doing things?'

Here is the game plan for foreign investors: Start with the customer. Promote alliances with the network of family connections and personal relationships. Offer investment deals that fit the local government's plan to grow the domestic economy and build market share abroad. To gain the trust of Chinese-owned firms, foreign firms must become dual citizens, too. Dual citizenship is the crucial cultural difference between the way business is carried out in China, Taiwan and the countries of South East Asia versus the West.

Cultural memory

South East Asia bestrides the crossroads of several important cultures. Here's what we know about the general patterns of behavior among the different communities who live in South East Asia.

- *Confucian dynamism.* Singaporeans have a long-term orientation.[4] They use traditional family loyalties to form networks of business interests among the ethnic Chinese throughout South East Asia, in the China commonwealth (Taiwan, Hong Kong and mainland China) and with Overseas Chinese in the US, Canada and Australia.
- *Islam.* Indonesia, peninsular Malaysia, the southern Philippines, and Brunei are linked to co-religious communities in South Asia and the Arab Middle East.
- *Buddhism.* Thailand combines elements of Hinduism and shares portions of this tradition with Sri Lanka and Cambodia. The Buddhism of China and Japan is different to the tradition found in Thailand.
- *Christianity.* The Philippines is primarily a Catholic country. Pockets of Christianity exist in all countries, especially among the ethnic Chinese,

and these reflect the religion of the former colonial power: Calvinism (the Dutch in Indonesia); Anglicanism (the British in Singapore and Malaysia); and Catholicism (the French in Indo-China).

- *Capitalism.* Dutch and British export-oriented practices, Japanese habits of governing the market, American preferences for freer business decision making, ethnic Chinese desire for family members to run the business, and Malay partiality for acceptable Moslem practices on interest payments and women at work: all these offer global firms a rich selection of choices to promote global business in South East Asia.

Is there a link between national culture and economic growth? Can Singapore's recent emergence as an economic 'tiger' be explained by Confucian dynamism, or the 'Protestant' work ethic of the ethnic Chinese, who constitute 85% of the island's population, and the local government's role in governing the market?

Networks of extended families

Chan Heng Chee, the director of the Singapore International Foundation, places culture (or Asian values) within Singapore's social reality. 'The values are universal, but the way those values are acted out in Asia is different. [She cites] "intense dependency on the extended family" ... [as] a social reality which is vital to the Chinese, and not just a desirable goal.'[5] Here are some definitions of culture:

- Culture is the entire way of life of a group of people (or social reality).
- Culture is learned behavior (about work, consumption, and savings).
- Cultural change is a switch in the way people do things (with new technology).
- Social change is recasting human relationships (within domestic and foreign firms).

In fact, Singapore reverses many basic ideas of Western (or American) capitalism.[6] For example, the economic performance of individuals must be placed within the social reality of the family. Group bonding includes ragsto-riches success by entrepreneurs within families. Second, business management includes values as an equal part of decision-making under uncertainty. Ritual and a deep respect for traditional authority are paramount within family firms. Third, family transactions come first; familial ties through extended networks come second; and deals with strangers come last. The best transactions are with family and friends. Fourth, ambiguity and discretion are inherently better than formal laws and predictability to encourage insiders to do a better job for the extended family and keep outsiders at bay.

Ethnic Chinese in South East Asia tend to have a major share of economic activity. Chinese populations are distributed as follows: Malaysia, 33% or 5 million; Indonesia, 5% or 7 million; the Philippines, 10% or 8 million; and Thailand 7% or 3.5 million. The numbers in the last two countries are underestimated. Many native-born citizens, who have been in both countries for 100 years or more, still live by Chinese cultural values. Since their business activities sweep across all sectors of the national economies of South East Asia, some of which are dominated by Islamic values, ethnic Chinese business executives must first unscrew their Chinese head when they seek business opportunities outside the extended family. Then they must screw on their local head when dealing with governments in Malaysia and Indonesia. Moreover, ethnic Chinese must screw on their foreign head when venturing into joint ventures with US and Japanese global firms.

Within families fathers tend to have only a Chinese head, and they value tangible assets such as real estate. They look for quick deals that give them cash. They operate on instinct, call all the shots, and value loyalty more than skills. The sons and a few daughters (many trained in the US) are faced with some confusion between personal friendships and business deals with non-family members. Since these second-generation managers must begin to work outside the close-knit world of overseas Chinese, they are less committed to tradition. As the years go by, the children will be at the heart of deploying the wealth of their parents and helping Chinese-speaking Asia attain its rightful place in the global economy. The children will be running very large conglomerates, a new cadre of global firms from South East Asia.

Win–lose win–lose world

The U-World is at the beginning of a different mix of trade and investment opportunities in South East Asia and its 'near abroad' neighbors. Vietnam could be a showcase for Asean investors as the region tries to stay close to its three most important markets – namely, Japan, China and the US. Cambodia and Laos could come along too in the wake of successes in Vietnam and Thailand, but foreign investors might prefer to put their money in China and Vietnam only. Burma might parlay the interest of the oil and gas industry in the country's reserves into an acceptable marriage with Asean and the US; however, it is more probable that US opposition to Burma's military rulers will propel Burma into Salmon Day.

Family-owned global firms

> *'Most of the new [Asean] multinationals are offspring of old family empires that dominate local economies. In Indonesia, Liem Sioe Liong's Salim group alone is estimated to account for 5% of Indonesian GNP. Other Chinese families run 17 of Indonesia's biggest 25 conglomerates, while Chinese–Thai business families have dominated the economy of Thailand for many years. Malaysian–Chinese families, such as the Lees and the Kuoks, have expanded rapidly despite government policies favouring Malay-owned businesses ... These local conglomerates in particular illustrate the new multinational mentality.*[7]

Let's look at a few examples of how these Asean-based global firms position their assets both within and outside South East Asia and Indo-China.

- Charoen Pokphand (CP) of Thailand, which is owned by the Chearavanont family, brews beer, produces petrochemicals, and builds motorbikes with Honda in China; farms shrimp in Mexico; and installs phones in Bangkok. Fifty percent of its assets are outside Thailand.
- Lippo of Indonesia, which is owned by the Riady family, has 40% of its assets outside of Indonesia, mostly in Hong Kong and China.

- The Keppel Group of Singapore, which is 33% owned by the government, produces half its profits from overseas, e.g. ship repair and financial services in the Philippines, and building modern office blocks in Vietnam. Even today only a small amount of their assets are in Vietnam, and almost none of them are in the rest of Indo-China and Burma. These Asean-based firms prefer tangible results from the China commonwealth and in other high-growth areas of the world.

Corporate assets

By the mid-1990s, these family-owned conglomerates begin to change their business tactics. They need professional managers who speak English. So they turn to Philippine managers, who are reputed to be the best in the region, to run the Indonesian banking sector, e.g. The Lippo Bank, First Pacific (Salim Group).[8] Then CP recruits American managers from Nynex who have skills in telecommunications, and Raja Garuda Mas (Indonesia) hires foreign managers to help start up or manage its businesses. Also CP, Lippo, Keppel, and Sime Darby (Malaysia) form joint ventures with global firms. Finally, when the founding Soeryadjaya family sells out local professional managers weather the crisis and keep the core business prosperous for Astra International of Indonesia.[9]

All of these changes in business tactics show the influence of the sons and daughters on the internationalization of family-owned conglomerates within South East Asia. We are just a few years away from major changes in the capital structure of and the practice of management in firms from Singapore, Malaysia, Indonesia, Thailand and the Philippines.

Networks of alliances

Foreign-owned global firms must understand and predict the general patterns of behavior of family-owned conglomerates from the countries of South East Asia.[10] This knowledge permits global firms to select local partners with desirable behavioral characteristics, and give them (and the local government) a share in ownership. Also this information helps global firms keep tighter control over joint ventures with local private firms and state enterprises.

What we can do with Asean-based, family owned firms?

- Singapore firms stress organization and discipline, promote transparency and the payment of local taxes, and demand politeness from foreign firms.
- Indonesia firms understand extended family obligations, and practice ritual politeness to family insiders and foreign outsiders.
- Malaysian and Indonesian firms assist government in promoting local native partners for foreign firms.
- Thai firms use shyness to promote friendliness with foreign firms.
- Philippine firms stress innovation and professionalism with foreign firms.

What do we get from the emerging global firms of South East Asia?

- Product differentiation encourages low price strategies (Indonesia and the Philippines).
- Technological capacity produces product adaptation for smaller markets (Singapore).
- Skill intensity keeps average wage payments competitive (Malaysia and Thailand).
- Absolute capital requirements permits the use of fungible capital throughout the region.
- Scale economies promotes sales by Japanese and American global firms.
- Domestic policy regime provides a key competitive advantage (Singapore and Malaysia).

Here's w*hat we are going to do on Monday that's different.*

- Know about differences in national culture and their impact on business alliances.
- Join forces with networks of extended families and friends.

- Take part in the boom enjoyed by the Association of South East Asian Nations (Asean).

Intra-regional trade

Intra-regional trade is significant for all members of Asean. With average per capita economic growth of 5% per year and average savings rates equaling 30% annually, the Asean countries are becoming go-go economies in which costs, prices, and interest rates all start going up. Without common central bank institutions for the region, the Asean countries must deal with current account deficits, shortages in infrastructure and skilled workers, and overheated economies on their own. Japanese and American investors find strong governments, extended family businesses, local partners, and ritual politeness as crucial to business success.

American and Japanese investments add to domestic savings, transfer improved technologies, create learning opportunities, and support internal economic growth. No market failure exists in these foreign investments. American and Japanese executives transfer their know-how to their government and private-sector partners. Also they accumulate experience and learning about the Singapore market, and adapt it for use in Malaysia, Indonesia, Thailand and the Philippines. With the active support of local governments, both sets of foreign investors create the conditions for product differentiation, and offer benefits from scale economies to the nation-states of Asean. They obtain capital for quality projects and to repatriate revenue back to Japan and the US.

Successful Japanese and American business executives understand the expected behavior of national groups, ethnic Chinese and foreign outsiders. They know to stress organization and discipline in Singapore, and to practice ritual politeness in Indonesia and Malaysia. These managers accept Thai friendliness as a product of a colonial-free past, and the Filipino interest in innovation as a reminder of America's control of the Philippines. Foreign business executives recognize the emergence of family-owned conglomerates into new global firms, and how the latter are changing their business tactics to meet the needs of home and host country markets. Looked at carefully we find that Asian managers have reversed many Western ideas about capitalism, and yet these executives are successful investors both at home and abroad.

Loyalists versus U-World

During Japan's rise to economic prominence, East Asian nations followed trade and investment rules set by the Europe and the US. Japan, Korea, China, and the countries of South East Asia were divided by geography, colonial conquest, inherited culture, religion, language and political 'isms'.

Today, economic integration, computer-based telecommunications, and the newest manufacturing technologies are propelling them into a more unified region of the world. The Overseas Chinese throughout South East Asia, their kin in the China commonwealth and their extended families in the US, Canada and Australia make the larger Asia–Pacific region more prosperous than the North Atlantic region.

Clearly, the West needs Asia's 500 million middle class. Their average savings rate is 30% annually, far higher than in the US and Europe. Asians aspire to all the consumer and industrial goods, and services recently found only in the West.[11]

Korea, Hong Kong, Taiwan and Singapore (the 'tigers'); Malaysia, Indonesia and Thailand (the 'tiger cubs'); the Philippines; and coastal and southern China are important internal markets in which governments force companies to beat world standards of performance. Governing the market in East Asia means learning how to cope successfully with accelerating technological and economic change. There is the West's opportunity for future business.

Economic trend spotting

Overall, South East Asia has strong economic fundamentals. Beginning in 1975, the Asean countries have averaged a GDP growth rate of around 6%. They are 'the world's foremost supplier of several major commodities producing and exporting more than 90 percent of the world's rubber, 70 percent of its tin and vegetable oils, 60 percent of sawn logs and almost 50 percent of sawn timber as well as substantial amounts of rice, sugar, crude petroleum, pineapple and spices'.[12] Here are some of the current economic problems facing the Asean countries.

Singapore

During 1995, 'manufacturing was the star performer, expanding output by a hefty 12.3 percent. Electronics, which contributes more than two-fifths of the country's industrial output, soared to 19 percent. Productivity jumped 8.6%, more than double the 3.9 percent increase in the overall economy.'[13] Although inflation in 1996 was a low 1.4%, manufacturing costs (especially for land, rent and wages) in Singapore keep rising. Some production is shifting across the Straits to Johore Baru in Malaysia.

Malaysia

Since manufacturing is dependent on the import of Japanese components priced in expensive yen, Malaysia's current account deficit is $7.1 billion (in US dollars), or 9% of GDP.[14] That's too high by world standards. However, the government will not lower economic growth targets to solve its economic problem. Instead, it is pushing big Malaysian groups – both Chinese and Malay-owned – to venture abroad through exports and foreign investments.

Indonesia

The country's current account deficit is $7.9 billion (in US dollars), or 3.8% of GDP.[15] Although that's OK by world standards, Indonesia is plagued by non-oil imports outstripping non-oil exports as its 190 million people (The country is the fourth most populous nation) race towards middle class prosperity. The fear is that the national economy will overheat, bringing with it double digit increases in prices. Nevertheless, the dominating fact about the islands of Indonesia is they are rich. They are the biggest emerging market of East Asia.

Thailand

Superficially, there is little to worry about: the economy is booming with full employment and short-term financial stability. Thailand offers foreign investors cheap labor and a pro-business government policy. The details show something different: infrastructure development is in chaos, skilled labor is in short supply, and traditional conservative monetary policy and strict fiscal discipline are at risk. The current account deficit is above 6% of GDP.[16] That's a bit too high. As a consequence of trade and financial problems, in late 1996 Moody's, the US credit-rating agency, downgraded Thailand's short-term sovereign debt ratings, and in April 1997, Moody's downgraded Thailand's long-term sovereign debt rating.[17] Thailand remains in a state of denial about its current economic problems.

The Philippines

Since the economy grew 5.1% in 1994, this South East Asian country was labeled 'a truly unexpected Asian economic miracle',[18] says Michael Taylor, an economist with Morgan Stanley. Then came the collapse of Mexico. Many investors saw similarities between the two countries, and thought the Philippines would become roadkill, too. It didn't work out that way. Both Clark and Subic Bay, two former American air and naval bases, are back in business as economic hubs. In fact, in 1995 exports grew by at least 6%, an excellent record, but they have suffered afterwards in the slowdown throughout the region.

However, unwise loans on real estate (where the property has gone through asset deflation) have pushed the banking industry throughout East Asia into near collapse.[19]

- *Japan*: 'Lenders have at least $216 billion in bad loans.'
- *China*: 'As much as 40%, or $240 billion of the state banks' loan portfolios are worthless.'
- *Taiwan*: 'Lenders have made few provisions against bad debt write-offs.'
- *Thailand*: 'A property bust has left Thai commercial banks with $15.5 billion (in US dollars) in a rising wave of nonperforming and bad loans.'
- *Indonesia*: 'Nearly 20% of the loans at seven state-owned banks, which represent half of the nation's banking assets, look shaky.'
- *South Korea*: 'Banks have $11 billion in bad loans.'

Success is no longer a sure thing even with family and personal connections. Governing the market in hopes of building up market share abroad – that is, the allocation of capital by bureaucrats – left price-sensitive commodity industries (such as memory chips, electronics and chemicals) overbuilt, overexposed and overdue. As a consequence, interest, which was pegged at rates well above Libor or world rates, is not being paid to local banks on projects for building new shipping terminals, airports, commercial buildings and apartment blocks. Therefore, local banks have had to sell their foreign investments at fire-sale prices to keep themselves afloat. A few might be bailed out by local governments; others could go under. Most will have no funds to fuel another round of investments in East Asia by local and foreign business firms.

Nevertheless, the money will come from the international financial community because the economic growth rates of most countries in East Asia will still be twice as fast as those in developed world of the US, Europe and Japan. For example, in 1996, even with knowledge of the banking crisis, private capital went to East and South Asia in record numbers:[20] China, $52 billion; Indonesia, $17.9 billion; Malaysia, $16 billion; Thailand, $13.3 billion; and India, $8.0 billion. If the countries of South East Asia are quick to adopt the best available technology across all sectors and they are able to convince their friends in the China commonwealth to abandon state subsidies of sunset industries, then the richer East Asian countries might overtake the rich West.[21] Any estimates of total GDP and per capita GDP for the region show coastal China, Taiwan, Korea, most countries in South East Asia and India growing substantially (but not miraculously) relative to the rest of the world today, tomorrow and well into the future.[22]

Trend spotting free trade

Asean is the door through which its seven members ask foreigners into the region. Let's tick off the numbers for investors.

Intra-regional Asean trade grew 40 percent to $111 billion (in US dollars) in 1994. 'The trend in Asean is the exchange of intermediate products where companies source in bulk. Asean is one market where investors can segment production and reap economies of scale,'[23] says Julius Cesar Parreñas of the Center for Research and Communication in Manila. For example, 'Thailand's partners in Asean [have replaced] the US as the country's leading export market in 1995, accounting for more than 22 percent of exports. Japan is not far behind, buying 17 percent of Thailand's exports.'[24]

Asean's target is to reach a unified common tariff of 5% on all goods by 2004, and to work towards open markets in financial services, aviation, communications, transportation, and construction. Asean's 400 million people will double their GDP to $1200 billion (in US dollars), perhaps as quickly as the turn of the century. Note: Asean still has a long way to go towards any harmonization of customs procedures and relief from nontariff barriers.

Today, Asean is America's fourth largest trading partner (behind Canada, Japan, and Mexico), and ranks third in Asia as a destination for US investment, behind Japan and Australia.[25] According to the US Department of Commerce, Asean will be the US's largest emerging market trading partner by the year 2000. Two regional trends – infrastructure improvement and increasing preference for a high-protein diet (for example, American-style hamburgers) – will increase imports from the US.[26]

Here are several case examples:[27]

- *General Electric* (US) opened a $20 million plant in Surabaya, Indonesia to supply lighting products to the entire region. A third of the output is destined for the Indonesia market and two-thirds for export, especially to Thailand and Malaysia. GE has six major manufacturing plants within Asean. Three are in Indonesia: a locomotive factory in Madium, central Java, and two lighting plants in Java. The others are a plastic-resin plant in Singapore, a factory making temperature controls in Malaysia and a plant that manufactures refrigerators in the Philippines.
- *General Motors* is building a $750 million plant in eastern Thailand to make a modestly priced car for export to the whole of Asia.
- *Ford* is building a $473 million plant in eastern Thailand to make pick-up trucks. Most Asean countries want a fully integrated car industry of their own, but the economics of auto assembly operations requires a regional approach.
- *Kellogg's* is building a $11.7 million plant in eastern Thailand to supply Asean with cereals from rice, wheat and maize.
- *Kraft's* sales of cheeses, powdered drinks, coffee and confectionery have grown by 15% a year since the early 1990s.
- *Wal-Mart* in an alliance with Lippo group opened its first store in Jakarta, Indonesia in late 1996.

Trend spotting management style

Although some influential Western observers see growing sameness among peoples in South East Asia, foreign investors who have to make deals with local families, executives and entrepreneurs note very specific differences in how managers approach business decisions. Let's do a checklist of the knowledge that foreign business executives are suppose to know before they can do successful deals in South East Asia.[28]

Singapore

- Stress organization and discipline.
- Be transparent and pay local taxes.

- Acknowledge impact of Confucian dynamism.

Malaysia and Indonesia

- Pay attention to government's interest in promoting local partners.
- Practice ritual politeness.
- Acknowledge impact of Islam.

Thailand

- Understand the cultural tradition of shyness.
- Value friendliness.
- Acknowledge impact of Buddhism.

The Philippines

- Encourage an innovative approach to life.
- Value competence.
- Acknowledge impact of Catholicism.

All Asean countries

- Be polite, don't lose your temper and never be rude.
- Understand extended family obligations.
- Value loyalty.

Antonio Lopez, a management consultant at the Asian Institute of Management in Manila puts it this way: 'When somebody asks me: "What is the Malaysian way of doing business?" I say: Do you mean Chinese Malaysian, Indian Malaysian or Malay Malaysian? Are we referring to Penang or Kuala Lumpur? I could go through the same routine with every country except Singapore.'[29]

South East Asia's big assignment

As the Asean nations seek to install a lower common tariff for all of South East Asia and Indo-China, they are watching carefully the emergence of a competing idea, free trade in goods, services, and capital among the 18 nations of the Asia–Pacific Economic Cooperation group. APEC includes three G7 countries (the US, Canada and Japan), China, Korea, six Asean countries, Australia, Mexico, Chile, and Papua New Guinea. These 18 countries account for half of world output.

'The conventional wisdom among western opinion leaders [is] that free markets and price stability are preconditions for viable economic development and comprise such elements as balanced budgets, monetary stability, pegged exchange rates, and privatization of state-owned enterprises.'[30] The nation-state provides the regulatory framework for private-sector transactions, and the world economy is open to free trade in goods, services and capital. This is the Washington Consensus.

International investors want the most efficient economy, the most convertible currency, the most educated work force, the sharpest business managers. The new conventional wisdom is to adopt the policies of the Washington Consensus – but simultaneously to build a robust financial, transportation, and energy infrastructure for corporations to compete effectively in domestic and global markets.[31] The world's capital markets prefer low-risk opportunities based on economic fundamentals and underlying investment values.

In East Asia, Japan practices selective adoption of Western technologies and competes for market share throughout the region. China overcomes a century of Western spheres of influence, but then accepts Western and Japanese foreign direct investments. And South East Asia recoils from Western colonization, and makes itself over to become a supplier to Western and Japanese markets.[32] Although East Asia is little more than a geographical expression, its national powerhouses force Western-oriented global business to rethink ideas about the positive role of government, family business relationships, and building social consensus. These national cultural differences are important knowledge for global firms.

East Asia is united in trade and investment, but lacks a center, a single all-important national currency, a dominant nation-state. The cultures of

Korea, Japan, China, Indochina, South East Asia and Australia are so different to form anything but the loosest political forum for discussion of common economic problems. Today, their common experience is being tied together with countries on the other side of the Pacific Ocean through largely autonomous networks of family-owned and private-sector global business firms.

According to Japan's foreign minister, most Asian members of APEC prefer the Japanese approach towards trade, or flexibility and harmony, over the American approach of tightly written rules and adversarial conflict.[33] 'In Asia we agree to things only if they are subject to change without any advance notice ... If we want the governments to endorse the plan, then it is important to make it clear from the beginning that they will not have to follow it,'[34] says Narongchai Akrasanee, chairman of General Finance & Securities, a brokerage company in Thailand. Member countries do not want APEC to become a free trade customs union (such as NAFTA, EU and Asean). Nor do the smaller members want to give up supporting key industries in which the unofficial rules of ritual politeness are more important than the WTO idea of transparency and openness. APEC is not based on a treaty, nor is it an action-oriented, US-type trade organization. Liberalization of trade is to be done 'flexibly,' a Japanese, Korean, and Chinese approach to solving trade problems among nations. In short, the APEC goal is to promote Asian harmony, an industrial policy heavily influenced by the approach of Japan's Ministry of International Trade and Investment (MITI) towards Japanese firms.

The U-World of financial capitalism

The Asean countries nurture the inflow of foreign capital by setting up rating agencies to assess the credit worthiness of domestic bond issues. Thailand's Rating and Information Services (TRIS), the Rating Agency of Malaysia (RAM), and Indonesia's Pefindo do the same work as America's Moody's or Standard & Poor's. Today, TRIS has an on-line information service with Bloomberg, an American provider of financial data.[35]

Within the global economy, the Asean countries cope with asymmetric shocks through national links to international capital markets. Each uses computer-based information technology to link national exchanges to those in Tokyo, New York, and London. Together, they are preparing South East Asia to play a larger role in equity deals, bond issues and derivatives. Success in building a financial infrastructure is a good test of their global business skills.

For example, Singapore's Simex is tied to the Chicago Mercantile Exchange (Merc) and the London International Financial Futures and Options Exchange (Liffe). This alliance shares products (such as bond futures and options), keeps exchange operations at capacity and reduces the costs of regulation. Malaysia sets up the Kuala Lumpur Options and Financial Futures Exchange (Kloffe) to compete with Singapore. According to Khairil Anuar Abdullah, a director at Malaysia's Securities Commission, Kloffe's 'stock index futures, options on the index, and options on individual stocks will increase the underlying liquidity of the cash market and allow investors to manage their portfolios and risks better'.[36]

The financial markets of both Singapore and Malaysia are being helped by the return of Hong Kong to mainland China. Singapore offers a strong national economy, large domestic savings, and an unrivaled infrastructure. Malaysia has the first two but lacks good financial infrastructure. To overcome this deficiency, Malaysia is offering foreign fund managers the government's Employees Provident Fund for investment in stocks; currently, Malaysia's private and corporate wealth is between $72 and $80 billion (in US dollars).[37] Singapore's counter offer is to permit foreign fund managers to manage $40 billion in assets held by the island's Central Provident Fund and the Government of Singapore Investment Corporation.

'Geography, rather than cost or political considerations, seems to matter most when companies decide on a regional base.' If fund managers and business executives are focusing on China, today the better choice is Hong Kong, but tomorrow it might be Shanghai. If financial executives are upgrading their position in South East Asia, the better choice is Singapore. 'Significantly, many banks and financial-services companies have moved some of their departments (mainly those involved in back-office functions) from Hong Kong to Singapore. This includes Swiss Bank, Royal Bank of Canada, American Express, J.P. Morgan, Merrill Lynch, Lehman Brothers – and even the crown colony's premier financial institution, Hongkong & Shanghai Banking Corp.'[38]

The results show up in the market for Asian currencies. National governments set free the Japanese yen, Malaysian dollar, Thai baht, and the Indonesian rupiah. And the market has found ways to get around government controls over the Singapore dollar, the South Korean won, the Chinese renminbi, and the Indian rupee.

Inflows of foreign exchange comfort investors because they know they can take their repatriated profits home. Sometimes, hot money crosses the exchanges and this speculation complicates monetary policy. During the Mexican peso crisis, all countries with large current account deficits had to do what the Bank of Thailand did, raise interest rates to prevent the baht from devaluation pressures. Now it can fall back on its bond repurchase or 'repo' agreements with other central banks in the region.

When the US Federal Reserve raises interest rates the flow of funds from the US to South East Asia begins to dry up. Also Japanese investors shift their flow of funds from the Asean countries to the US. Such a shift in funds flow is called a flight to quality in global capital markets. Only Singapore stands to benefit among the Asean countries.

When and whether the fundamental strengths of the Asean countries are going to have a chance to show through again is the open question facing the markets in Singapore, Malaysia, Thailand and Indonesia. Asean current-account deficits differ from Mexico's in two ways. 'The first is that they largely reflect Asian countries' hefty imports of capital goods; Mexicans, in contrast, have been importing relatively more consumer goods. The second is that the flows of foreign capital needed to finance these current-account deficits have tended to take the form of more enduring foreign direct investment, rather than volatile portfolio flows.'[39] Judging the attraction of emerging markets in South East Asia means looking as much at likely perceptions among investors as at the economic and management fundamentals.

Let us draw these conclusions:[40] first, 'no matter how strong the economic fundamentals of an emerging market, it can still be affected by events on the far side of the world'. Second, 'predictability of central-bank action is crucial to effective competition for an often volatile pool of international capital'. Third, 'stable economic policies [must] reassure domestic investors, who will provide the bulk of investment capital'.

These are the strengths found in some of the Asean countries.

Win–win world

The U-World is at the end of the beginning of a new win–win world market era for Japanese firms. These direct investors come to South East Asia with several comparative advantages – that is, public policy skills in working with government, a capability to manufacture value added technology appropriate to local costs and better sourcing arrangements. Japanese parent firms transfer some of their knowledge to subsidiaries in South East Asia. The former's competitive advantage is they link in-process work from South East Asia to final production in Japan, and both to high-tech markets in the US and Europe.

Since 1992 nearly a fifth of Japanese foreign direct investment has gone to Asia. Moreover, Japan's Export–Import Bank predicts that, between 1996 and 1998, three-quarters of Japan's new foreign direct investment will go to Asia, much of it to China and South East Asia.[41] Why are Japanese firms investing so heavily in Asia? Perhaps, because the integration of Asia is proceeding with much greater ease and much less fuss than other attempts to create regional blocs, NAFTA and EU. Here are their reasons for investing in South East Asia.

Labor costs

Haruo Tsuji, president of Sharp, the electronics company, says: 'Labour costs in Japan [are] the highest in the world ... All operating costs, such as the cost of manufacturing sites, construction, water, electricity, and shipping, are also the highest in the world.'[42] Compare these costs with those in the South East Asia countries. According to the 8th Survey of Japanese-affiliated Manufacturers in Asean, monthly wages paid to plant workers average US$221 in Malaysia, $215 in Thailand, $191 in the Philippines, and $115 in Indonesia.[43]

Local sourcing

Some 39.8% of the materials and parts used by these subsidiaries of Japanese manufacturers are purchased locally; 42.9% comes from Japan; and 20% comes form other sources. Increased purchasing from local sources goes on year after year. These are just three examples.[44] Japan's Sanyo Electric Co. uses 50% local content to produce refrigerators in west Java, Indonesia. Thai ball bearings roll out from Japan's supply network into the dollar bloc of Asia. And Hitachi Consumer Products Malaysia in Selangor produces the chassis (including the circuit board) for projection TVs sold in the US.

Intra-Asia sales

Of the 452 Japanese firms participating in the Survey, 83.3% export products to Japan (40.4%), other Asean markets (18.8%), and the Asian newly industrialized economies (13.9%), such as Korea, Taiwan, and Hong Kong.[45]

Internal demand

'The growth in demand for Japanese products is particularly strong in ... South East Asia and China. If demand is to be met in these growth markets, Japanese manufacturers need to be located in them to ensure that they are in close touch with local needs and trends.'[46] The flotsam of Japanese culture is scattered about South East Asia: *karaoke* bars in Bangkok; canned iced tea in Singapore; 'Oshin', a Japanese TV series about a long-suffering, hard-working woman, on Star TV translated into Mandarin, Cantonese, and Fukien Chinese; and *pachinko* (pinball-style machines) parlors in Taiwan. Japan's Asian neighbors want Japan's cultural exports.[47]

Currency issues

Is Japan's enthusiasm for South East Asia a bubble? Possibly. Are there limits to how many factories the Japanese will build in cheap Asian sites to make basic products? Yes, but the stock of Japanese investment in Asia will still grow. Will Asian firms denominate more of their trade in yen, especially to Japanese partners? This depends on the dollar. If it falls in value in terms of the yen, those South East Asian countries that peg their currencies to the dollar and operate at full capacity will experience inflation (Singapore, Malaysia, and Thailand); and those that are poorer will see a spurt in demand for their exports (Indonesia, the Philippines). If the dollar rises in value (which means the yen falls), then the South East Asia countries will find it much easier to pay back their yen denominated debt. However, the cost of their exports will rise and this means Japan must invest elsewhere – in China, India or other lower-cost Asian countries.

All Japanese firms have gone through an economic revolution in which asset deflation has led to early managerial retirements and employee lay-offs at home, and additional investments and new jobs in South East Asia. The companies are strong competitors. Their weakness is as follows: Japanese firms prefer to join forces with other Japanese firms, and thus they are outside the family connections and personal relationships so important to success in South East Asia.

Networks, passports and Indonesia

American and Japanese firms compete simultaneously in several South East Asian markets. They and their suppliers and customers are affected by changes in public policies. 'Governments shape the pattern of foreign investment not only directly, through their own investments (in which case state enterprises may "crowd out" foreign firms), but also through the regulatory and licensing regime. This is particularly so in developing economies, where distrust of multinational corporations … has been a powerful factor in the postcolonial environment of many countries.'[48]

Let's take a look at the policies of Indonesia. 'First, from 1974 until 1992 a local joint venture partner was required, and in a number of very large investment projects [such as power and telecommunications] that partner was the central government. [Second], independently of these joint ventures, the state has been a significant industrial investor, in sectors where there has been a potential "crowding out" effect on [global firms. Third], for much of the 1970s and 1980s there was a highly interventionist trade regime, and 'tariff bargaining' as an inducement to invest was widespread. Finally, certain industries were closed entirely to foreign equity investments.'[49]

In the oil and financial industries, America is the largest investor in Indonesia. Also Indonesia's Lippo Group has formed joint ventures with car maker Chrysler, retailers J.C. Penney and Wal-Mart, power-plant builders Misson Energy and Entergy, and financial-services company First Union. Moreover, the US Export–Import Bank lent $168 million (in US dollars) to Bakrie Brothers, Indonesia's largest *pribumi* (native Indonesian) business group, to set up a fixed wireless communications system in co-operation with government-owned Telkom.[50] Finally, Indonesian private-sector conglomerates, which are mostly owned by ethnic Chinese families, are using joint ventures with foreign investors to become more market-oriented and accountable to the public.

What kind of competitive advantage?

What are the competitive advantages of foreign-owned firms? Here is a consensus about the factors for profit maximization among foreign direct investors, and their application to Indonesia.[51]

- *Product differentiation* depends on the availability of advertising to promote brand names and the willingness of foreign firms to carry out low price strategies in markets in which incomes are low by world standards. In the case of Indonesian manufacturing, Japan has been the dominant foreign investor, reaching a high of 75% of total foreign direct investments in the 1980s.
- *Technological capacity* requires product adaptation to smaller markets, reduced prices, and assembly and processing skills. On the other hand in the case of Indonesia, large-scale projects that are eligible for project finance need little technological adaptation when the partners agree to government suggested prices.
- *Skill intensity* suggests global firms have first-hand experiences in many different countries. In Indonesia, labor markets are competitive and average wage payments are lower than elsewhere in South East Asia.
- *Absolute capital requirements* means foreign firms are closely tied to inter-

national capital markets, and the former can invest both international and domestic funds in whatever sectors are opened by local governments. Indonesia competes with other Asean countries for fungible capital from international sources.

- *Scale economies* give foreign firms a competitive advantage over domestic firms in technology and export orientation. Indonesia needs both and actively recruits both Japanese and American foreign investors through liberalization efforts of the its national government.
- *The domestic policy regime* helps larger state enterprises crowd out global firms. In Indonesia, the former needs the latter because MNEs have technology, sourcing ties, and overseas customer markets crucial to the local government.

Two paradigms

The Washington Consensus is the paradigm of economic growth and development preferred by the US and Europe. It commands funds from the World Bank, the US Export–Import Bank and the international financial markets. The Washington Consensus is America's view on how the global economy should function within a fully deregulated, *laissez-faire* environment favored by Washington, DC, the Thatcherite revolutionaries in London, England, and the Anglo-American establishment of economists from Stanford, Chicago, MIT and Oxford. The best policy for all countries consists of joining the Washington Consensus so that the owners and managers of international capital are not upset by deviations from free trade norms.

Of course, the East Asian miracle did not flow from this pragmatic approach towards economic growth and development. Most Asians from East and South Asia hold different views. For example, Japan's plan-rationale approach towards its domestic economy and external markets is *not* part of the paradigm. The government–business ties of Taiwan and Korea to govern the market through support of sectoral industrial policies is *not* part of the paradigm. Singapore's high savings rate and commitment to advanced education are *not* part of the paradigm. The commitments of the South East Asian countries and the China commonwealth to extended family connections and personal relationships are *not* part of the paradigm.

Therefore, East Asia's growth and development paradigm underscores economic and social structures within nation-states rather than erring on the side of markets set up to encourage private transactions in all sectors of the economy. The intellectual culture of Japan, the China commonwealth, South East Asia and India is at odds with the currents of thought dominating the US, the UK and elsewhere in the West. Failure to pay attention to these real differences in intellectual thought and culture practices means the West and its global firms might find themselves shut out from crucial investment projects in East Asia.

Costs and benefits

How should the countries of South East Asia weigh the costs of attempting to deepen their commitments to the global, largely American economy against their preferences for the plan-rationale approach towards the domestic economy, intraregional trade and external markets in the US, Japan, China and Europe?

This is a case of decision-making under uncertainty. If Asean continues with its gradual reduction of intraregional tariffs, APEC, which includes virtually all countries bordering on the Asia–Pacific basin, might overtake the budding free trade initiative of the South East Asian countries. If Asean staples itself to Japan, the South East Asian countries tie their trade and investment options to a slow growth, developed consumer market with whom they share views about strategies for economic growth and development. If Asean staples itself to China, the South East Asian countries tie their foreign trade options to a fast growth, developing consumer market in which the connections of kith and kin play a crucial role in success or failure within a socialist economy with market characteristics. If Asean accelerates its reduction of intraregional tariffs, APEC might be left behind as the US promotes free trade between NAFTA and AFTA (Asean Free Trade Area). However, if free trade between the US and Asean starts up too soon or goes too far too fast, Singapore, Thailand, Indonesia and the Philippines might line up against Malaysia, Vietnam, Japan and China.

This possible division of Asean members and of East Asian countries goes against the preference for harmony among Asian nation-states.

Decision theory cannot create certainty where none exists, but it offers a way to organize one's thoughts and gather the information needed to make better decisions. Consider reimaging South East Asia. The best policy for the Asean countries is to use these next few years to sort out whether they want to go slowly within the larger Asian context or strike out boldly in a new direction with the US and, by extension, Europe.

Remember the South East Asian countries have options. One option is to do things gradually so that harmony reigns among all Asean members. This will 'Keep the memory of the past in the present'. This mastery of drift fits well into Confucianism and slips nicely into the more conservative habits of Islam.

Option two is to make a commitment for Japan. Many political leaders in South East Asia are trying to forget the excesses of World War II, and they are trying out new forms of economic co-operation with Japan. They tend to be sympathetic with the way in which Japan governs the market, and most have tried out various aspects of the plan-rationale approach within their domestic economies. Also Japanese political leaders are walking very slowly into a new era of trade and investment relationships with the Asean countries. Both sets of leaders want to 'Recognize their own mistakes and gather the courage to correct them'.

Option three is to make a commitment for China. A few South East Asian countries might be comfortable within a greater China commonwealth. Singapore has guided an investment strategy for Shanghai, and Thailand has sent its family-owned, private-sector firms into coastal and inland China. Malaysia and Indonesia let its overseas ethnic Chinese invest in mainland China, but these Malay-dominated countries will not be restful within a greater China commonwealth. All South East Asian political leaders prefer the known of Asean to a leap into an unknown socialist future with market characteristics. These leaders 'Expect their colleagues to correct their mistakes'.

Option four is to accelerate the shift to free trade and crown its success with a deal between the US and Asean. Yes more money might flow from the World Bank, the US Export–Import Bank and international capital markets. But at what price? Division among member countries of Asean, and unhappiness on the part of both China and Japan.

Today, the South East Asian countries are pursuing option one. Will they be able to do this successfully? Yes. During the next few years, all Asean members will maintain their agreement that this option is the best one for the region.

Decision theory offers us a way to reflect on our choices and make the best decisions we can with the information we have in front of us. Decision theory offers South East Asia a way to organize itself to avoid even the appearance of Salmon Day.

Salmon Day

Salmon Day is the experience of spending an entire day swimming upstream only to get screwed in the end. In South East Asia, American, Japanese and locally owned firms are fighting the global fight, gaining market share and winning support on Wall Street. None of these firms were victims of Salmon Day because they invested within the success stories of Asean. However, if they put their investments in Asean's 'near abroad', these firms could face Salmon Day.

Here are some summary thoughts. First, the transmitters of South East Asia's common knowledge about intraregional free trade offer the following: participation in the East Asian, largely Chinese and Malay life styles with the possibility that parts of both of them might become an American life style as well; English and non-Mandarin forms of Chinese as languages of communication with the first having the primacy of use in the short run; Mandarin Chinese as the language of communication and having the primacy of use in the long run; and South East Asia's big assignment – namely, continued opening of the region's national economies towards intraregional free trade.

Second, the editors of South East Asia's common knowledge are in the process of reimaging their intellectual thoughts, conclusions and policy prescriptions concerning the U-World and the world of Asian loyalist nation-states. Confucianism, Buddhism, Islam and Christianity are the strong intellectual currents, and they compete for popular support against Western ideas. The costs and benefits of doing a better job of additional globalization among family-owned businesses depends on the ability of parents to given their adult children a freer hand in running these emerging global firms and on the quality of foreign partners from the US and Japan.

Third, managers and others in the professional class must decide whether South East Asia uses the plan-rationale or the market-rationale approach to expand free trade. Both paradigms help governments move closer towards more intraregional free trade. If South East Asia decides to staple itself to the US, Asean gains more access to crucial export markets and avoids Salmon Day. If South East Asia decides to staple itself to Japan or China, Asean gains more access to other crucial export markets, and could avoid Salmon Day. None of these decisions force a dangerous swim upstream and the possibility of getting screwed in the end.

South East Asia is better off than eastern Europe, Central Asia, Latin America and sub-Saharan Africa. This region can choose its destiny rather than be tied to Germany, Russia, the US or Europe. Although the Philippines has had its ups and downs over the last decade, none of the original six members of Asean are failed nation-states. And a more recent member, Vietnam, is well on its way to stronger ties with the international economy. Unfortunately with Burma, Laos and Cambodia as members, Asean might have a Belarus, Mexico or Sierra Leone among the roadkill of the global economy.

Wisdom: decision theory lays out possible choices. Pay attention: none of the choices lead to Salmon Day, but some are better than others. Open the doors: South East Asia makes a choice to expand intraregional free trade slowly and in harmony with its neighbors in the Asia–Pacific Basin.

What do I get?

Here is our to do list:

- Buy into South East Asia's big assignment for more intraregional trade.
- Accept the US, Japan and China as crucial to success in South East Asia.
- Build on the memories of Confucianism, Islam and the market system.
- Reimage ideas about family networks and personal connections.
- Recognize rooted cosmopolitans as twenty-first century citizens of South

East Asia.

- Create a greater China commonwealth and pan Asia–Pacific stream of business.
- Use decision theory to weigh costs and benefits of options.

Remember these words: Salmon Day is not in the cards for South East Asia. Its political leaders will have to work very hard to screw things up and force their nation-states into Salmon Day.

Successful decisions

A successful East Asia makes the following policy decisions:

- East Asia works with America to create an Asia–Pacific free trade area.
- China, Japan and South East Asia work together to create free trade in the region.
- Japan and South East Asia help China join the WTO.
- East Asia gives limited assistance to South and Central Asia, and Russian Far East.

Conclusions

'Freer trade is ultimately beneficial ... but a victory for globalization that comes at the price of social disintegration will be very hollow indeed.'[1]

Dani Rodrik, Harvard University.

'No issue displays more facile unanimity among editorial writers than the virtues of trade and the selfish ignorance of protectionists.'[2]

Robert Kuttner, coeditor of *The American Prospect.*

Free trade in a global world

Who needs globalization anyway? Wouldn't the world be better off without the uninhibited pursuit of unfettered markets? Don't they worsen the consequences of economic competitiveness because nation-states are unwilling to make sacrifices for the common good? And if all *laissez faire* markets disappear, with their know-it-all attitudes about free trade, would anyone really miss them?

Yes, some would be missed. Here's the argument:

- Memory mothers internationalism and xenophobia, babies globalization and nationalism, and nurses the interaction and misunderstandings of governments, business firms and citizens with markets and nation-states. Free trade's big assignment is to absorb both the good and bad in international relations, and to help leaders avoid the stigma of governing failed nation-states.

- Americans believe in American exceptionalism, and that the US is unique among nations. Free trade's big assignment is to ease Canadians, Mexicans and others into accepting America's view that free trade within North America depends on adversarial free enterprise capitalism.
- Europeans practice free trade within an emerging Franco-German political structure that is alien to the US, disagreeable to the UK and suspect to many within Europe. The EU forces an answer to the question: 'So are we Europeans or aren't we?' Free trade's big assignment is show western Europe how to bring eastern Europe into a more wholesome relationship with the larger North Atlantic community.
- Managing global business is about judging the certainties of tomorrow's NAFTA, EU, Asean and APEC within both the market-rationale and plan-rationale approaches towards the international economy. Also managing global business is about judging the certainty of China's realpolitik, its balancing of market forces and socialist practices, and its forthcoming decisions concerning Japan and the US. Free trade's big assignment is to work through the anti-NAFTA and anti-EU phobias of the 1990s, and to tie together the global prospects of the North Atlantic and Asia–Pacific communities for the twenty-first century.
- Divining what connects Japan and China and what each means to the other is not within the memory of most Americans and Europeans. Free trade's big assignment is to encourage the West to understand how dual, identity and network citizenship of family and personal relationships; and national political commitments create business success in East Asia and might be appropriate for some in North America and Europe, too.

Therefore, free trade's overall big assignment is to put nation-states in a 'golden straight-jacket' so they can export products, attract foreign investors and become global competitors. (This is a term used by the British to explain how Mrs Thatcher's economic policies changed the UK for good.) Free trade forces governments to practise sound budget and fiscal policies so their sovereign debt qualifies for favorable ratings in the world's bond markets. Free trade encourages governments to practice competent industrial policies so their state enterprise sector can be made profitable, privatized or sold to foreigners. Free trade suggests governments tax less for social welfare projects so more money is left with private interests to invest in capital improvements and employee retraining. If done well, free trade helps nation-states avoid Salmon Day.

Literate criticisms

Not everyone agrees with these modest proposals for free trade within the global marketplace. The other side argues that governments must reclaim their power from free trade markets. First came the Americans Ross Perot and Patrick J. Buchanan with arguments about how economic integration (code word NAFTA) was impoverishing America. Of course, these proved to be specious as America grew jobs, expanded exports, and invested in Ontario and northern Mexico. Free trade cannot lift all of Mexico out of its endemic poverty; it cannot quickly make eastern Europe as market-oriented as western Europe; and it cannot turn all of East Asia into prosperous little Singapores within any time frame useful to business executives. These are a bridge too far for free trade.

Later Sir James Goldsmith, the Franco-English European, spoke out against giving up national power to faceless foreign bureaucrats (code word EU). Of course, Brussels issued fewer regulations as the European nation-states reclaimed some of their legislative powers in their quest for co-operative ways to enhance Europe and resolve several important outstanding issues, such as monetary union. Free trade can cut a path through the thicket of state enterprise capitalism. However, free trade cannot force all countries to march in unison down this road; cultural differences require each nation-state to go at its own pace towards the highway of the future. This bridge with its many on and off ramps is worth building within the broader North Atlantic community.

Now George Soros argues governments are giving up their social safety nets for the poor to placate currency markets, investment houses and international funds. Of course, rich government-managed safety nets are out of fashion in the US, in doubt in the UK and open to question in Europe. On the other hand, family connections are the favored safety nets in East Asia. Free trade offers the West an inoffensive way to learn a thing or two from how East Asians deal with the problems of the old, sick and the poor, failed corporations and long-term unemployment among managers and workers. This is an appropriate extension of the bridge between the North Atlantic and East Asian communities.

Let's be clear. Free trade is not a panacea. Some nation-states succeed within the market-rationale approach towards economic growth. Others find their history, culture and society demand the plan-rationale approach towards the domestic economy and its ties to the global market. Free trade offers the world a future of possibilities about how to organize economic activity.

Globaloney

Such meticulous written exercises are not new, where what passes for pontificating about globalization is really globaloney about economic integration. The dream world of NAFTA extended to all of the Caribbean and Latin America loves platitudes more than it loves hard facts. US politicians can get attention by throwing up their hands and blaming mysterious global forces for fictitious job losses among low skilled, unemployed Americans. That's why the quiet growth of jobs in the American mid-west where products are manufactured for sale in the Canadian province of Ontario and in northern Mexico is without cheerleaders. That's why globalization's effects have been understated by many Washington-based economists because the real benefits of free trade are maddeningly hard to measure.

Their efforts at quantifying the capitalist threat is at best computer-enhanced popcorn, a debasing of critical economic thought by those sniping business executives turned writers, who look down their noses at the rest of us because they think we are not on to their game. What do they want? Opportunities to sow doubt and make money with a contrarian view of markets. Obtaining the ear of politicians in Washington and London so that free trade will not be able to drive towards its future without detours. Gaining a cabinet position and an administrative slot with one of the world's institutions for acolytes of the writers of globaloney.

Those who pontificate sneak into our consciousness with their immodest proposals about taming the excesses of a global free market. Who would jettison Mexico and leave America's southern frontier exposed? Who would close Europe to some or all of eastern Europe? Who would banish the Turks and other Moslems to permanent second place within Europe? Who would fail to grasp Japan's concern for its future without recognizing China's importance to East Asia? Who would refuse South East Asia's desire to be the fourth part of the developed world? Who would tell India, Iran, Central Asia, the Arab states of the Middle East and North Africa, sub-Saharan Africa and the Caribbean that free trade, global markets and capitalism are not for them? No-one wants to be left out of the global market. This is the policy problem facing bureaucrats and business executives around the world. They are not going to be easy to resolve.

U-World versus loyalists

Clearly, the newest members of the U-World empire have struck back. The West is faced with that increasingly visible phenomenon whereby many strong competitors for world market share are coming from East Asian business executives, who are using the marketing strategies they've learned from Americans and turning them on their head. These 'Chinese' are both distant and close, outsiders and insiders; they inject something new into ongoing business organizations. Yet we have two colliding worlds that fail to understand each other today, but had better make a stab at it tomorrow.

Although it is hard to know just what we may need on this trip to a new century, it is time to think seriously about packing. Shall we pack for the real world of loyal nation-states or the ideal U-World? Shall we carry an APEC bag, check in a NAFTA suitcase or have others haul a EU steamer trunk?

Will a simple tick of the clock in 2000 or 2001, whichever we take to define the millennium, ease the fearsome sting that the global, largely American life style has long conveyed to Canadians, Mexicans, Europeans west and east, Japanese, Chinese and other East Asians? Globalization, after all, will bear almost a messianic relation to the next century as exports and imports (and trade deficits) have held for the twentieth century. Yet America's unwavering preference for the market-rationale approach seems more likely to lose its urgency in the wake of Japan's modification of the plan-rationale approach for the purpose of adding the views of China and South East Asia into the world mix. Unfortunately, the plan-rationale approach still awaits a broad appreciation in the West, although it does not lack for spirited advocates.

Are we going to be able to live with East Asia's plan-rationale approach intersecting with the West's market-rationale approach? The honest answer is that we don't know.

Vivid memories

Good intentions got roads paved. Until the mid-1990s, Americans were envious of Japan, Inc. and the juggernaut industrial economy called the EU. If you did not mind where the market-rationale road led, it was fun to walk the trade and investment road into Canada, Chile, Poland and Thailand. However, there came a time when destination Mexico, Ukraine or Vietnam started to look less inviting. Looking abroad, Americans were caught up in self-criticism about whether they should have heeded helpful hints from East Asian friends about the positive qualities of the plan-rationale approach towards growing the economy.

Turning back was the obvious solution for America, at least if one was traveling without Canada, Mexico, Europe, Japan and East Asia. For all the wealth of the American free enterprise system, America seemed pitiful and helpless when compared to German thrusts into England, eastern Europe and southern South America; and Japanese conquests of markets in China, South East Asia and in the US itself. Americans thought they could turn to emulation. They tried to govern the market with the globaloney of strategic trade theory. American politicians even tried to reimage free trade and NAFTA as a successful venture of America's new interest in government and business working together to make America competitive again.

However, in a crowd of successful politicians, their ever eager bureaucrats and financial supporters, and their friends from the executive suite, this forced march down a new road was not so easy, especially when it entailed dismantling the road as well. In addition, most of the free enterprise crowd pointed out that the former destination is exactly where they wanted to go, and no group of plan-rationale advocates was going to be allowed to spoil the capitalist party.

The essential argument of governing the market was that Anglo-Americans, who could not read Japanese and Chinese government and business documents in their original languages, had misunderstood the governing process in Japan, the China commonwealth and South East Asia. The plan-rationale approach was originally thought to be the creation of the Japanese, but from the early 1990s the role of intellectual creator passed irrevocably to the Mandarin Chinese. Business executives, especially in the US though not confined to it, constructed an image of their ideal competitors as successful organization men (preferably Japanese salarymen), owing nothing to the family connections and personal relationships of the rest of East Asia. In so doing, they suppressed the truth that most of the Japanese convoy system was derived from China and the overseas ethnic Chinese, in order to bolster up a view of Japan as the natural leader among East Asians against the communists who ruled China. Much of China's efforts was reduced to a level of a puppet on a string. A self-possessed culture such as the US–Japan consumer revolution was inclined to believe that it alone had the solutions, while impoverished China, India, Central Asia and Russia had gone down blind allies.

Even this would have attracted little attention if it had not been for the fact that China was reimaging its socialist system with market characteristics, pushing Shanghai and coastal China to the forefront of world commerce and trade, and that the overseas Chinese were doing in and out of China what no combination of Japanese and Americans could do by themselves. The exact relationship between a few Anglo-Americans pushing the plan-rationale approach and the emergence of the overseas Chinese in the ranks of global enterprise is unclear, and in a sense it hardly matters. What is important is the end result. Governing the market grew up among Japanese, Chinese, South East Asians and a few Anglo-Americans, and they challenged the paradigm in place at the World Bank. Decades official indifference, or misrepresentation in Washington, DC, have suppressed this fact.

The central thesis is untenable. The plan-rationale approach developed in East Asia is one important paradigm that the world lives with in many places outside of North America. The market-rationale approach is another important paradigm, especially for North America and in some parts of Europe, too. Both should be admired for what they are – that is, a way of organizing economic society for domestic growth and foreign sales. They imitate the histories, societies and cultures of different nation-states. Throughout the post-World War II period, both have been exposed to the stimulating efforts of the other and this explains some of the odd resemblances between them.

Reimaging globalization

America's mood has flipped. Triumph and self-satisfaction are the rising American themes. US global executives demand emulation for generating wealth and prosperity in the highly competitive global economy. The mantra of the market-rationale approach is '*laissez faire*, deregulate, downsize, cut taxes and do not interfere with the market.' Is this success cyclical? Will Japan or Europe or both rise again and reverse the roles? Does America really have a better formula for doing business?

So governing the market is not dead. The East Asian approach to the economy is also at the center of contemporary concerns. It too shaped the global economy. Since the Anglo-Americans failed to keep them in this place, we should pause for a moment before going for broke with the market-rationale approach. One of the lessons for politicians and bureaucrats in Washington and London is that intentions have consequences, and these may be the opposite of what was intended. Roads like these are inviting to many who wish to maintain the established paradigm at the World Bank, and we need all the challengers from East Asia and the West we can get, if we are not to continue down them at the expense of the alternative paradigm, the plan-rationale approach with its equivalent record of successes and failures in growing the economy.

Perhaps we will never solve the problem of competing approaches towards economic growth. Are both orthodox and catholic – that is, intellectual ideas that proclaim universal truths? Is one orthodox and the other heresy? But which is which? Might we reject both of them tomorrow as past mistakes? More than likely we will only get bored with them.

Trend spotting

Our subject is globalization. It is a much used and frequently abused term. The concept settled so well in our vocabulary that we speak as if its meaning were clear and obvious to everyone, and to everybody in the same way. Yet to spell out a definition of globalization with which all agree without a long string of qualifications is difficult. Globalization is a market-based economy, and that means a place in which private-sector transactions occur without too much government interference. Such economy lives in chaos, it itself is the chaos which strives to give itself a form, but a form never fixed for all time. This rampant dynamism of the global economy derives from the independence of firms on the one hand, and the prudent interventions by governments on the other. Theirs is a marriage of convenience, joined together by mutual dependence, and a struggle over rights and responsibilities without resolution.

At its best, the market snaps up extra resources and directs them to where they are best used, without any intervening mechanism. Should these successful enterprises always survive? Both the market-rationale and the plan-rationale approaches say 'yes'. At its worst, the market rejects those whose marginal productivity is too low to allow them to purchase the capital for future investments. Should these failed enterprises always perish? The market-rationale approach says 'yes', but the plan-rationale approach equivocates, says 'maybe yes, maybe no'. In Europe, Mexico and Latin America, and China and Korea, failing state supported enterprises were allowed to go on longer than necessary. These state enterprises took too many resources, crowded out worthwhile private initiatives and put a drag on the national economy. They were unable to compete in the emerging global market economy. All over the world governments took the initiative and began privatizing or closing failing state enterprises. Here too we have an example of the plan-rationale's approach towards governing the market.

Governing the market calls for government to encourage technological change, offer incentives for more investment capital and then induce firms to offer products at competitive prices both at home and abroad. Globalization succeeds when both government and markets play their most appropriate roles. This is the new paradigm for tomorrow's bureaucrats and business executives.

America, Europe, East Asia

Now that it is clear America will not subscribe to the plan-rationale approach, it is the job of East Asia and Europe to produce a new economic paradigm. It is here that governments and business executives must splice myths and roles together in what we might call feats of enterprise engineering.

On these grounds, Japanese global firms, especially their transplants in the US, have made a genuine contribution to business culture. What their cyclical successes and failures really show is that Japanese global firms are squarely in both the plan-rationale and market-rationale tradition of aggressive engineering and superior marketing. This is a role also being played by other canny, go-for-broke foreign investors like German global firms.

But would there be so many armchair critics if global success weren't accompanied by the onrush of family-owned ethnic Chinese firms from the networks of personal connections of South East Asia and China? Or the other global tribes from India, the Moslem states, Israel and the ethnic and religious minorities in Russia? Would the critics complain if global success were still confined within the Anglo-American, northern European and Japanese communities?

Business executives carry the nation-state's dreams of mega-success on their bottom lines. As unrecognized millionaires, they get away with doing deals over oil rights with whatever government controls the fields, pumping stations and pipelines. Yes they pay lip service to the US government's predilections for economic sanctions, but they let their Canadian subsidiaries go in and out of Cuba, and their German subsidiaries do business with Iran.

In choosing to go along or wink at American prohibitions, these executives give a fresh spin to the classic modernist dictum that governing the market means governments actually do something good, bad or indifferent for the domestic economy. Japan needs oil so much that no Japanese government would think of banning oil from Iran even if it were pumped and made fungible in the ports of the Arab states of the Persian Gulf. For mainstream trade talks with the US, where both governments want less noise and more agreement, Japan dons its most appropriate *kabuki* costumes – a walking history of Japan's long history of making incremental adjustments to forces outside its control. This genuine reimaging of Japan should be carefully followed by America, China and the countries of South East Asia because it may be the portent of the future. As Japanese history tells us: hidden decisions are better than public decisions or no decisions at all.

Furthermore, the Chinese are bringing some fresh touches to the notion of Asian business executives. According to the 1970s definition, Asian business executives produce cheap foreign goods and sell (or dump) them in the West. But by the 1990s a different kind of Asian business executives has emerged. Now they have studied and absorbed certain crucial, daring elements of global business strategy. Think Unilever, P&G and Kao. Then look at Tiger Balm from Singapore, or Norinco from China.

On the one hand, family-owned ethnic Chinese firms are determined to flaunt their new found success, aggressively defending every action, unwilling to admit their local governments have ever been wrong about societal values, personal connections and family ties. In short, these firms are a stereotype for Indian and Moslem enterprises that seek entry into the global market. For now, anyway, their success as role models for American firms depends on their ability to sell world-class goods in global markets.

Salmon Day redux

The book *Salmon Day* is about the decline, fall and rise of globalization. American exceptionalism was just a pile of bric-a-brac – meticulously piled up and saved, a heap of nothing. American uniqueness was killed off during a protracted, undefined economic upheaval in domestic and foreign places. The Europeans were obsessed by a single question to the exclusion of everything: what will happen to this economic union of ours? Will we achieve success? Will we be copied by the US and East Asia? The Japanese punctured their comfortable ideas as rational economic beings in a fit of cyclical demise and asset deflation. The Chinese yanked the rug out from under their own socialist system and labeled it, of all things, socialism with market characteristics. The overseas ethnic Chinese of South East Asia are contemporary Everymen – knowledgeable, capable and close-mouthed. They are former students of the MBA who went uphill from there.

How important are the ethnic Chinese to the overall scheme of things? This is a painful question for Americans, Europeans and Japanese. The Chinese are flowing almost seamlessly into the global marketplace. They are examples of more to come from the Indians, Arabs, Africans and others who right now are outside the global, largely American world market. The Chinese possess a sensibility to the possibilities of globalization that is, yes, extra special. They will offer the West a morally ambiguous judgment about what it takes to make a deal. They will show the West how utterly beyond definitive interpretation are its laws, administrative regulations and corporate acts when these come up against a government willing to use its powers to extend its reach across the Asia–Pacific region.

The Chinese grab for global market share is becoming, in fact, more and more a sort of postmodern plan-rationale approach towards the market – one very much of our time and about our own business lives. Their game plan is a self-help manual whose multinational ties let our imagination stretch to all points of the compass. The Chinese might sacrifice one set of personal relationships with government officials to family connections, but never both to the outside demands of foreign-owned global firms. The Japanese did the same when it came time to close ranks within the *keiretsu* against American-owned multinationals. Nothing is predetermined because all enter the global market with a different set of cultural habits and expectations about future success. Thus does China's grab for global market share become postmodern.

At stake, again and again, is the fragility of understanding, the troubled future of the two paradigms. The problem China poses is how the global economy can continue in a world ruled by petty party bureaucrats, excessive price cutting by failing state enterprises and a great deal more supply than demand. It may be that these economic problems are a prelude meant to be transcended, and we must be the subtle interpreters of the future. The story of China points forward toward a promise of prosperity even in the hinterlands of East Asia.

Successful firms avoid the bullet of Salmon Day while unsuccessful firms are roadkill on the road towards globalization, the real victims of Salmon Day. The following is a summary of the cases discussed in the book.

America

Mobil Oil, Monsanto, Owens Corning and Budweiser fought the global fight, gained market share and won support on Wall Street. Smaller US manufacturers took their newest technologies, applied them in a high-wage country and still made money. However, R.R. Donnelley, Whirlpool, Maytag and Pepsi fought the global fight, and lost market share to their competitors. Only Maytag turned itself around to the satisfaction of Wall Street. In Mexico, Bell Atlantic and Iusacell failed, and Telmex gained market share in the cellphone segment of the Mexican telephone market. Only Panamco, a Mexican-owned bottler for Coca Cola, won support on Wall Street for its Mexican and Latin American investments.

Europe

Interbrew and Daimler–Benz did very well overseas, increased share in market segments and niches, and won support in the City (London). Many American, some Japanese and a few German firms invested in Britain and gained market share in Europe. They took advantage of lower UK labor costs that came about because Britain was outside the ERM, Social Chapter and the emerging Euro. All foreign investors in central Europe sought market share without a certainty that Poland, the Czech Republic and Hungary would become members of the EU. These investors bet on the come. All foreign investors in Russia sought market share without a full and complete set of information about costs, sales and profits. Western-style accounting reports are not available. All foreign investors bet on the come in central and eastern Europe, Russia and the former Soviet Union without any certainty that the economic hopes might become market realities.

East Asia

Softbank, Kyocera and Sony gained market share and won support in Tokyo and on Wall Street. Motorola, Lucent and Microsoft fought for market share in China; they are doing what 'old China hands' from the US did unsuccessfully in the 1980s, and what Japanese investors, on the whole, have done well in the last two decades. In South East Asia, American, Japanese and locally owned firms took on each other for dominance in the Asia–Pacific region.

Global, largely American life style

Government and business agree that America is the only country that is both a political nationalizer and economic globalizer. The US uses its nationalism to strengthen the global, largely American life style in Europe, Japan, China and East Asia. This makes the US unique among the nation-states of the world. Canada and Mexico are neither political nationalizers or economic globalizers. They cannot use their nationalisms to strengthen their position around the world. This makes them a part of the larger US economy that is now known as NAFTA.

All leaders concur the nation-states of Europe (east and west) have been strong political nationalizers, especially, the Franco-German partnership that dominates crucial decisions about the future of the EU. On the other hand, the European Common Market–Community–Union is an important emerging economic globalizer that uses European 'nationalism' to strengthen its global position within central and eastern Europe.

Trend-setters recognize Japan, China and the countries of South East Asia are essentially political nationalizers. Japan uses nationalism to strengthen its global position within China and South East Asia. China uses dual, identity and network citizenship to strengthen family connections and personal relationships among the peoples within the China commonwealth. South East Asia prefers neither to be a political nationalizer or a economic globalizer. Instead, the Asean region wants to maintain its ties to the US, Japan and China so that it stays ahead of eastern Europe, Central Asia, Latin America and sub-Saharan Africa.

A winning Monday strategy

Here's how business firms from any place in the world can compete effectively in global markets. It's a short list of what American, Canadian, European, Japanese and ethnic Chinese firms are doing already.[3]

- *Mine demographic data.* Do a great deal of research on why things are and whether anything can be changed. Know a country's history, cultural and societal ambitions. Find out how consumer lifestyles, incomes and age distribution are changing.
- *Position products within local market segments.* Adapt products, packaging and promotion from home markets to the preferences of customers in host markets. Know a people's routine social practices before investing in factories, distributors and advertising agencies.
- *Team up with local partners.* Marry a local partner with good family connections and personal relationships. Know who can resolve employee and currency problems, and who can make a joint venture work.
- *Answer the 'What do I get?' question for customers.* Target products to fit local expectations about price, quality and value. Know how to exceed the expectations of local customers; they will come back for more and increase market share.

If you think global firms are special because they pal around with world governments and invest in all interesting global markets while mid-size and smaller firms stay closer to home, think again. For when all is said and done, what does New York, London or Tokyo – the crucial capitals of world capitalism that look down on the rest of the world – really consist of? They have executive egos, financial support systems and diabolically subversive ways to undercut competitors and governments alike in the quest for market share, profits and returns on investment. However, as one Japanese business executive recently pointed out, no industry remains profitable after 40–50 years of uninterrupted success. Thus governments must be on the lookout for the next success stories. Otherwise, their nation-states will join the backwater of others that failed to change with the times.

In summary, governments and business firms must work together to help nation-states avoid Salmon Day. Bureaucrats and business executives must start out on Monday to implement the steps listed above or face Salmon Day.

Endnotes

Introduction

1. John R. Walter. 'How to Leap Before You Look: Lessons for a Digital World', *The New York Times*, November 10, 1996, p. 12F. This article was taken from a larger essay published in September 1996 in 'The C.E.O. Series' of the Center for the Study of American Business at Washington University in St Louis.
2. 'A Survey of the World Economy: The Hitchhiker's Guide to Cybernomics', *The Economist*, September 28, 1996, Survey, p. 3.
3. Walter, p. 12F.
4. Walter, p. 12F.
5. Walter, p. 12F.
6. Richard A. Melcher, 'Strategies: The Smudgy Legacy AT&T's New Prez Left Behind', *Business Week*, February 3, 1997, p. 36.
7. Melcher, p. 36.
8. Melcher, p. 36.
9. 'Born to see; meant to look', *Forbes*, March 10, 1997, p. 127. Patrick Smith, *Japan. A Reinterpretation*, (New York: Pantheon Books, 1997).
10. 'Modern times, old trends', *The Economist*, April 20, 1996, p. 68. The article is a summary of three studies on the similarities between today's global economy and that of a century ago. Jeffrey Williamson, 'The Evolution of Global Labor Markets since 1830: Background Evidence and Hypotheses', *Explanations in Economic History*, (1995). Williamson, 'Globalisation, Convergence, and History', *Journal of Economic History*, (1996). Kevin O'Rourke, Alan Taylor, and Jeffrey Williamson, 'Factor Price Convergence in the Late Nineteenth Century', *International Economic Review*, (1996).

11. Paul Krugman, *Pop Internationalism*, (Cambridge, Massachusetts: MIT Press, 1996).
12. Paul Krugman, *Peddling Prosperity*, (New York: W.W. Norton & Company, 1994), p. 257.
13. 'A Survey ...', *The Economist*, September 28, 1996, p. 4.
14. Krugman, p. 257.
15. Murray Weidenbaum, *Neoisolationism and Global Realities* (St Louis, MI: Center for the Study of American Business, Policy Study Number 130, May 1996), pp. 1–2.
16. Tony Jackson, 'Management – Intellectual Revolution', *Financial Times*, March 10, 1997, p. 22.
17. 'In praise of Davos Man', *The Economist*, February 1, 1997, p. 18.
18. Jackson, p. 2.
19. Robert Lenzner and Stephen S. Johnson, 'Seeing things as they really are', *Forbes*, March 10, 1997, p. 123.
20. John S. Reed, 'Oh, the Modern World!' *The CEO Series*, (St Louis, Missouri: Center for the Study of American Business, CEO Series Issue No. 7, August 1996), p. 2.
21. Ronald Henkoff, 'Growing Your Company: Five Ways to Do It Right', *Fortune*, November 25, 1996, p. 84.
22. Fred R. Bleakley, 'Multinational Firms Spent $325 Billion in 1995 on Foreign Direct Investment', *The Wall Street Journal*, June 5, 1996, p. A2.
23. Fred R. Bleakley, 'Foreign Investment by Multinationals Grew 40% in 1995, Lifted by Mergers', *The Wall Street Journal*, September 25, 1995, p. A2.
24. *UNCTAD, Division of Transnational Corporations and Investment.* Cited by Bleakley, June 5, 1996, p. A2. The 1996 figures come from David Wessel, 'Capital Flow to Developing Nations Surges 20%', *Wall Street Journal*, March 24, 1997, p. B9.
25. Bleakley, June 5, 1996, p. A2.
26. Bleakley, September 25, 1996, p. A12.
27. Bleakley, September 25, 1996, p. A12.
28. Douglas Lamont, *Winning Worldwide: Strategies for Dominating Global Markets*, (Homewood, Illinois: Business One Irwin, aka Irwin Professional Publishing, 1991), p. 13.
29. O.B. Hardison Jr., *Disappearing Through the Skylight: Culture and Technology in the 20th Century* (New York: Viking Press, 1989).
30. 'Peter Drucker's 1990s: The futures that have already happened', *The Economist*, October 21, 1989, pp. 19–20 and 24.
31. Marcia Berss, 'Whirlpool's Bloody Nose', *Forbes*, March 11, 1996, pp. 90 and 92.

32. Berss, p. 92.
33. Bruce Upbin, 'Global, schmobal', *Forbes*, March 10, 1997, pp. 64 and 66.
34. Douglas Lamont, *Winning Worldwide: Strategies for Dominating Global Markets*, second edition, (Oxford, UK: Capstone Publishing Ltd, 1996), pp. 249–50.
35. Patricia Sellers, 'How Coke is Kicking Pepsi's Can', *Fortune*, October 28, 1996, pp. 70–84.
36. Sellers, pp. 70–84.
37. Guy de Jonquières, 'Cheap labour loses its allure for investors', *Financial Times*, July 15, 1996, p. 3.
38. Neil A. Martin, 'Travelin' Man', *Barron's*, March 18, 1996, p. 16.
39. Martin, p. 18.
40. Sellers, pp. 70–84.
41. 'Global investing: Advantage, Coke', *The Economist*, November 9, 1996, p. 91.
42. 'The world that changed the machine', *The Economist*, March 30, 1996, pp. 63–4.
43. 'Ford has a long haul at Mazda', *Business Week*, October 7, 1996, pp. 108–14.
44. Deepak Lal and H. Myint, *The Political Economy of Poverty, Equity and Growth*, (Oxford, UK: Clarendon Press, 1996). Cited by Martin Wolf, 'The politics of growth', *Financial Times*, August 27, 1996, p. 10.
45. 'Are the poor different', *The Economist*, April 27, 1996, p. 86.
46. 'Globalisation: Fine for some', *The Economist*, September 28, 1996, p. 72.
47. Mancur Olson, Jr., 'Big Bills Left on the Sidewalk: Why Some Nations are Rich and Others Poor', *Journal of Economic Perspectives* (Spring 1996). Cited by Martin Wolf, 'The Poverty of Nations', *Financial Times*, August 20, 1996, p. 10.
48. Bernard Lewis, *The Middle East* (New York: Simon & Schuster, 1996). Cited by 'Cultural explanations: the man in the Baghdad café', *The Economist*, November 9, 1996, p. 23.
49. 'Cultural explanations ...', p. 24.

Chapter 1 American Heartland

1. Gary Taylor, *Cultural Selection*, (New York: Basic Books, 1996).
2. 'Land of the big', *The Economist*, December 21, 1996, p. 28.
3. Daniel Goleman, 'Anthropology Group Takes Activist Stand to Protect Cultures', *The New York Times*, March 19, 1996, p. B8.
4. Michael Specter, 'World, Wide, Web: 3 English Words', *The New York Times*,

April 14, 1996, p. 1 E. David Crystal, *English as a Global Language* (Cambridge University Press, forthcoming); cited in 'The coming global tongue', *The Economist*, December 21, 1996, p. 76.

5. Cullen Murphy, 'The Spirit of Cotonou', *The Atlantic Monthly*, January 1997, p. 15.
6. Tony Jackson, 'Global competitiveness observed from an unfamiliar angle', *Financial Times*, November 21, 1996, p. 18.
7. 'The oil buccaneer', *The Economist*, November 30, 1996, p. 72.
8. 'The oil buccaneer', p. 72.
9. 'The oil buccaneer', p. 72.
10. 'The oil buccaneer', p. 72.
11. 'The oil buccaneer', p. 72.
12. 'The green gene giant', *The Economist*, April 26, 1997, p. 66.
13. Thomas A. Stewart, 'Owens Corning Back From the Dead', *Fortune*, May 26, 1997, pp. 118–26.
14. Gene Bylinsky, 'Five Heroes of U.S. Manufacturing', *Fortune*, May 26, 1997, pp. 104 [A–H].
15. Robert D. Kaplan, 'Fort Leavenworth and the Eclipse of Nationhood', *The Atlantic Monthly*, September 1996, pp. 75–6.
16. Kaplan, p. 81.
17. Michael Moynihan, *The Coming American Renaissance: How to Benefit from America's Economic Resurgence*, (New York: Simon & Schuster, 1996).
18. Clyde Prestowitz, *Trading Places: How We Allowed Japan to Take the Lead*, (New York: Basic Books, Inc., 1988). Lester Thurow, *The Future of Capitalism: How Today's Economic Forces Shape Tomorrow's World*, (New York: William Morrow & Company, 1996). Paul Kennedy, *The Rise and Fall of Great Powers*, (New York: HarperCollins, 1988).
19. 'Land of the big', *The Economist*, December 21, 1996, p. 28.
20. 'The coming global tongue', *The Economist*, December 21, 1996, pp. 75–8.

Chapter 2 Nafta and Mercosur

1. Richard W. Stevenson, 'Nafta's Impact on Jobs Has Been Slight, Study Says', *The New York Times*, December 19, 1996, p. C1.
2. Michael M. Phillips, 'South American Trade Pact Is Under Fire', *The Wall Street Journal*, October 23, 1996, p. A2.
3. Anthony DePalma, 'Canada Mmmm, Beaver Tail', *The New York Times*, January 26, 1997, p. E3.
4. Sam Dillon, 'Mexico Watch It: Hot Stuff', *The New York Times*, January 26, 1997, p. E3.

5. Sam Dillon, 'Miracle of Manhattitlan: A Town Transformed', *The New York Times*, December 25, 1996, p. A4.
6. Thomas Sowell, *Migrations and Cultures: A World View*, (New York: Basic Books, 1996).
7. 'Mexico: A Mexican Morass for Bell Atlantic', *Business Week*, January 20, 1997, p. 42.
8. 'The Border', *Business Week*, May 12, 1997, pp. 64–74. 'Mexico's new frontier', *The Economist*, February 8, 1997, p. 41.
9. Gregory Rodriguez, *The Emerging Latino Middle Class*, (Los Angeles: Pepperdine University, 1996); cited in 'Latinos in California: The next Italians', *The Economist*, December 14, 1996, p. 29.
10. Richard W. Stevenson, 'Nafta's Impact on Jobs Has Been Slight, Study Says', *The New York Times*, December 19, 1996, p. C1.
11. Stevenson, p. C3.
12. James Goldsmith, *The Trap*, (London: Macmillan, 1994).
13. Robert Z. Lawrence, *Single World, Divided Nations?* (Washington and Paris: Brookings Institution/OECD, 1996).
14. Michael Sandel, *Democracy's Discontent: America in Search of a Public Philosophy*, (Cambridge, Massachusetts: Harvard University Press, 1996).
15. Gary C. Hufbauer, Institute for International Economics. Cited in 'Singing the NAFTA Blues', *Business Week*, December 9, 1996, p. 55.
16. Michael M. Phillips, 'South American Trade Pact Is Under Fire', *The Wall Street Journal*, October 23, 1996, p. A2.
17. Michael M. Phillips, 'South American Trade Pact Is Under Fire', *The Wall Street Journal*, October 23, 1996, p. A2.
18. Michael Reid, 'A Survey of Mercosur: Remapping South America', *The Economist*, October 12, 1996, Survey, pp. 3, 4 and 7.
19. 'Latin America's backlash', *The Economist*, November 30, 1996, p. 15.
20. 'The backlash in Latin America: Gestures against reform', *The Economist*, November 30, 1996, p. 21.
21. Michael Pettis, 'The Liquidity Trap: Latin America's Free-Market Past', *Foreign Affairs*, 75:6 (November/December 1996): 2–7.
22. Pettis, p. 6.
23. 'The myth of the powerless state', *The Economist*, October 7, 1995, pp. 15–16.

Chapter 3 The West

1. Edward Rothstein, 'Trend-Spotting: It's All the Rage', *The New York Times*, December 29, 1996, p. 1H.

2. 'Asians at Play: A good day out', *The Economist*, December 21, 1996, p. 47.
3. Anthony DePalma, 'Space, the TV Frontier, Now', *The New York Times*, December 30, 1996, p. C1.
4. DePalma, p. C1.
5. DePalma, p. C2.
6. Edward Rothstein, 'Trend-Spotting: It's All the Rage', *The New York Times*, December 29, 1996, p. 1H.
7. Rothstein, p. 28H
8. Wolfgang Saxon, 'Earl W. Count, 97, Chronicler of Christmas's Roots, Is Dead', *The New York Times*, p. 18Y.
9. Richard Pipes, *Russia under the Old Regime*, 2nd edn (New York: Penguin Books, 1995), p. 223.
10. Alessandra Stanley, 'Dec. 25 in Russia: The Adoration of the Monetary', *The New York Times*, December 24, 1996, p. A4.
11. David Remnick, *Resurrection: The Struggle for a New Russia*, (New York: Random House, 1997).
12. 'Russia: Unequal abroad, punchier at home', *The Economist*, March 22, 1997, p. 61.
13. 'Extracting oil from the Caspian: Great game, awful risks', *The Economist*, February 15, 1997, pp. 63–4.
14. Ahmed Rashid, 'Central Asia: Power Play', *Far Eastern Economic Review*, April 10, 1997, pp. 22–4.
15. 'Ready to Burn Rubber in Russia', *Business Week*, March 31, 1997, pp. 52–3.
16. Sophia Kishkovsky and Elizabeth Williamson, 'Second-Class Comrades No More: Women Stoke Russia's Start-Up Boom', *The Wall Street Journal*, January 30, 1997, p. A12. 'Update: A Newsletter for Investors in Russia', *The U.S. Russia Investment Fund*, 3:1 (Winter 1997).
17. John Thornhill, 'Analysts grapple with Russian valuations', *Financial Times*, January 31, 1997, p. 24.
18. The data come from 'Asians at Play: A good day out', *The Economist*, December 21, 1996, p. 47.
19. 'Asians at Play: A good day out', *The Economist*, December 21, 1996, p. 47.
20. 'Asians at Play: A good day out', *The Economist*, December 21, 1996, p. 47.

Chapter 4 Businessmen's Europe

1. Michael Sandel, *Democracy's Discontent: America in Search of a Public Philosophy* (Cambridge, MA: Harvard University Press, 1996), p. 339. Cites Shirley Williams, 'Sovereignty and Accountability in the European Community', in

Robert O. Keohane and Stanley Hoffman, eds, *The New European Community*, (Boulder: Westview Press, 1991), pp. 155–76.

2. 'L'avenir c'est moi', *The Economist*, January 11, 1997, p. 64.
3. This is the first line of André Aciman, *Out of Egypt*, (New York: Riverhead Books, 1994).
4. Glenn Collins, 'The Americanization of Salsa', *The New York Times*, January 9, 1997, p. C1.
5. Collins, p. C1.
6. Youssef M. Ibrahim, 'Finland: An Unlikely Home Base for Universal Use of Technology', *The New York Times*, January 20, 1997, p. A1.
7. Matrix Information and Directory Services in Austin, Texas. Cited in Ibrahim, p. C6.
8. Ibrahim, p. C6.
9. Roger Cohen, 'Lacking Barricades France Is in a Funk', *The New York Times*, December 29, 1996, p. E5.
10. Cohen, p. E5.
11. Martin du Bois, 'Interbrew's Toast: Here's to Niche Beers', *The Wall Street Journal*, November 26, 1996, p. A18.
12. Thomas L. Friedman, 'Europe, Not Japan, Is Called America's Frontier for Trade', *The New York Times*, June 9, 1996, p. C1.
13. 'Europe: Old World, New Investment', *Business Week*, October 7, 1996, p. 50.
14. The data come from 'European Business: Le Défi Américain, again', *The Economist*, July 13, 1996, pp. 21–3.
15. The idea comes from Lisa Jardine, *Worldly Goods: A New History of the Renaissance* (New York: Nan A. Talese/Doublelday, 1996).
16. 'Management theory: The advent of the Euroguru', *The Economist*, March 11, 1995, pp. 66–7.
17. 'L'avenir c'est moi', p. 64.
18. 'L'avenir c'est moi', p. 64.
19. Haig Simonian, 'Wheel turns full circle at Mercedes-Benz', *Financial Times*, January 20, 1997, p. 19.
20. Barry Eichengreen, 'A Handshake Unwise for Europe', *The New York Times*, April 28, 1996, p. F11.
21. Thomas Kamm and Cacile Rohwedder, 'Europe Quakes Anew at Germany's Incredible Bulk', *The Wall Street Journal*, January 9, 1997, p. A7.
22. Thomas Kamm, 'Snobbery: The Latest Hitch in Unifying Europe', *The Wall Street Journal*, November 6, 1996, p. A17.
23. Emma Tucker, 'No appetite to change the mix', *Financial Times*, October 31, 1996, p. 11.
24. Lionel Barber, 'EU's north-south divide narrows', *Financial Times*, November 7, 1996, p. 4.

Chapter 5 Euro-Realism

1. Rudi Dornbusch, 'Euro Fantasies', *Foreign Affairs*, 75:5 (September/October 1996), p. 124.
2. George Soros, 'Can Europe Work?' *Foreign Affairs*, 75:5 (September/October 1996), p. 11.
3. Paul De Grauwe, 'Towards EMU without the EMS', *Economic Policy*, 18 (April 1994), p. 173.
4. Peter Norman, 'Paris designs on EU bank attacked', *Financial Times*, January 20, 1997, p. 1.
5. John Kauffman, *At Home in the Universe: The Search for Laws of Self-Organization and Complexity* (New York: Oxford University Press, 1995).
6. 'The challenge to EMU: Europe learns its alphabet', *The Economist*, December 9, 1995, p. 20.
7. 'The etiquette of merging currencies', *The Economist*, December 9, 1995, p. 80.
8. 'The challenge . . .', p. 20.
9. Nathaniel C. Nash, 'Showing Europe that U.S. Still Cares', *The New York Times*, December 3, 1995, p. 12.
10. De Grauwe, p. 173.
11. Dornbusch, p. 124.
12. Soros, p. 11.The data comes from Edward Carr, 'Business in Europe: A fortress against change', *The Economist*, November 23, 1996, Survey, p. 3; and 'Redesigning the German model', *The Economist*, January 27, 1996, p. 41.
13. 'Murdoch's New Deal', *The New Yorker*, May 12, 1997, p. 38.
14. These comments were made by Lawrence Lindsey, a member of the US Federal Reserve's Board of Governors. Amity Shales, 'Loving the Mark', *The New Yorker*, April 28 and May 5, 1997, p. 193.
15. Shales, p. 188.
16. David Currie, 'The Pros and Cons of EMU', (London: The Economist Intelligence Unit, 1997). Cited in 'A little EMU enlightenment', *The Economist*, February 22, 1997, p. 26.
17. John Gray, 'The Tasks Ahead', *The Times Literary Supplement*, May 9, 1997, p. 13.
18. Gillian Tett, 'Reaping the benefits of devaluation', *Financial Times*, October 6, 1995, Survey: World Economy and Finance, p. 24.
19. Tett, p. 24.
20. Tett, p. 24.
21. 'A much devalued theory', *The Economist*, January 20, 1996, p. 71.
22. Tett, p. 24.
23. Tett, p. 24.

24. Michael Lindemann, 'German "job exporting" debate is renewed', *Financial Times*, November 2, 1995, p. 4.
25. Lindemann, p. 4.
26. Karen Lowry Miller, 'Siemens Shapes Up', *Business Week*, May 1, 1995, p. 53. Greg Steinmetz, 'New Siemens Strategy: Investing in Itself', *The Wall Street Journal*, February 14, 1995, p. A12.
27. Kevin Done, 'Average 4.4% GDP increase forecast for E Europe states', *Financial Times*, November 2, 1995, p. 1. The 1997 estimate is a composite figure from several sources.
28. Audrey Choi, 'For Mercedes, Going Global Means Being Less German', *The Wall Street Journal*, April 27, 1995, p. B4.
29. Choi, p. B4.
30. Choi, p. B4.
31. Peter Gumbel and Audrey Choi, 'Germany Making Comeback, with Daimler in the Lead', *The Wall Street Journal*, April 7, 1995, p. A6.
32. Julie Wolf, 'Maytag Unit's Jobs Transfer Stirs Ire in EC', *The Wall Street Journal*, February 1, 1993, p. B6B. Craig Forman, 'France is Preparing to Battle Britain Over Flight of Jobs Across the Channel', *The Wall Street Journal*, February 3, 1993, p. A17.
33. David Goodhart, 'Social dumping: hardly an open and shut case', *Financial Times*, February 4, 1993, p. 2.
34. Edward Carr, 'Business in Europe: A fortress against change', *The Economist*, November 23, 1996, Survey, p. 8.
35. Tony Jackson, 'Footloose across Europe frontiers', *Financial Times*, March 9, 1993, p. 15.
36. David Marsh, 'An elusive corporate consensus', *Financial Times*, February 24, 1994, p. 10.
37. Marsh, p. 10.
38. William Dawkins, Martin Dickson, and John Burton, 'Why a buoyant US and Asia pity poor old listless Europe', *Financial Times*, February 24, 1994, p. 11.
39. Dawkins et al., p. 11.
40. Clive Cookson and Andrew Fisher, 'Ideas people at back of the field', *Financial Times*, March 2, 1994, p. 9.
41. Cookson and Fisher, p. 9.
42. The examples were drawn from the following articles: 'Privatisation', *The Economist*', August 21, 1993, pp. 18–20; 'Europe for Sale', *Business Week*, July 19, 1993, pp. 38–39; and Charles Fleming, 'France Pinpoints 4 State Companies For Privatization', *The Wall Street Journal*, July 22, 1993, p. A11.
43. Martin Wolf, 'A relapse into Eurosclerosis', *Financial Times*, February 24, 1994, p. 11.
44. Andrew Fisher, 'Cost of being safety-conscious', *Financial Times*, December

2/3 1995, p. 8.
45. Richard Waters, 'Global expansion to cap it all', *Financial Times*, March 23, 1994, p. 14.
46. 'GE Capital's Grand Tour of Europe', *Business Week*, October 16, 1995, p. 62.
47. Peter Koenig, 'If Europe's dead, why is GE investing billions there?' *Fortune*, September 9, 1996, p. 116.
48. Koenig, p. 114.
49. M. L Pradelles de Latou, 'Identity as a Complex Network', in C. Fried, ed., *Minorities, Community and Identity*, (Berlin 1983), p. 79. Cited by Eric Hobsbawm in his Barry Amiel and Norman Melburn Trust Lecture, (London: Institute of Education), May 2, 1996.
50. Peter Doyle, John Saunders, and Veronica Wong, 'Competition in Global Markets: A Case Study of American and Japanese Competition in the British Market', *Journal of International Business Studies*, 23:3 (Third Quarter, 1992), p. 440.
51. Will Hutton, 'Relaunching Western Economies', *Foreign Affairs*, 75:6 (November/December 1996), p. 10.
52. Hutton, p.12.
53. Robert Medley, 'Keeping Monetary Union on Track', *Foreign Affairs*, 75:6 (November/December 1996), pp. 25–6.
54. Adam Gopnik, 'Elvis of the Élysée, *The New Yorker*, June 3, 1996, pp. 44–5.

Chapter 6 The East

1. Valéry Giscard d'Estaing, 'Europe: Les Raisons de l'Echec' ('Europe: The Reasons for Failure'), *Le Figaro*, January 10, 1995, p. 6. Cited in Robin Niblett, 'The European Disunion: Competing Visions of Integration', *The Washington Quarterly*, 20:1 (Winter 1997), p. 91.
2. John Major, 'Europe: A Future that Works', speech given at the University of Leiden, Netherlands, September 7, 1994. Cited in Niblett, p. 102.
3. J.G.A. Pocock, 'What Do We Mean by Europe', *The Wilson Quarterly*, 21:1 (Winter 1997), p. 27.
4. Pocock, pp. 17 and 28.
5. Michael Howard, 'Land of War, Land of Peace', *The Wilson Quarterly*, 21:1 (Winter 1997), pp. 34–5.
6. John Major, 'Europe: A Future that Works', speech given at the University of Leiden, Netherlands, September 7, 1994. Cited in Niblett, p. 102.
7. 'At Issue: Imagine, Europe', *The Wilson Quarterly*, 21:1 (Winter 1997), p. 9.
8. 'At Issue …', p. 10.
9. The phrase comes from Milovan Djilas, *The New Class*, 1957.

10. Peter Passel, 'Economic Scene', *The New York Times,* May 16, 1996, p. C2.
11. Passel, p. C2.
12. Michael Mandelbaum of Johns Hopkins made this deft comment. See Thomas L. Friedman, 'NATO or Tomato?' *The New York Times,* January 22, 1997, p. A19.
13. Barbara Beck, 'Germany: Divided Still', *The Economist,* November 9, 1996, Survey, p. 3.
14. Peter Norman, 'Saxony enters aid minefield', *Financial Times,* September 2, 1996, p. 10.
15. Beck, Survey, p. 5.
16. Beck, Survey, pp. 12 and 22.
17. Steve Liesman, 'The Outlook', *The Wall Street Journal,* July 22, 1996, p. A1.
18. 'Piling into Central Europe', *Business Week,* July 1, 1996, p. 42.
19. Daniel Michaels, 'Tentative `Tiger', *The Wall Street Journal,* November 25, 1996, p. A1.
20. Michaels, p. A1.
21. Source is Gazeta Wyborcza. Cited by Jane Perlez, 'A Bourgeoisie Blooms and Goes Shopping', *The New York Times,* May 14, 1996, p. C1.
22. Perlez, 'A Bourgeoisie ...', p. C8.
23. 'Piling ...', p. 42.
24. 'Piling ...', p. 43.
25. Michaels, p. A1.
26. Daniel Michaels, 'Auto Makers Beat a Path to Poland for Production, Sales and EU Backdoor', *The Wall Street Journal,* December 19, 1996, p. A15.
27. Neil King Jr, 'Czechs offer Peek at New Corporate East', *The Wall Street Journal,* May 8, 1996, p. A10.
28. Dean Calbreath, 'Czech Banks Buy Stakes in Biggest Clients', *The Wall Street Journal,* July 5, 1996, p. A4.
29. 'What's the Trade-In on a Trabi?' *Business Week,* July 15, 1996, p. 47.
30. Kevin Done, 'A drive into the fast lane', *Financial Times,* February 28, 1996, p. 11.
31. Jane Perlez, 'Shaking Up Czech Industry', *The New York Times,* January 6, 1966, p. 19.
32. Perlez, 'Shaking ...',p. 19.
33. Jane Perlez, 'Little to Spend, but Hungarians Like Mall', *The New York Times,* December 24, 1996, pp. C1 and C4.
34. Andrew Fisher, 'German banks offer an umbrella to E Europe', *Financial Times,* August 13, 1996, p. 15.
35. Fisher, p. 15.
36. Virginia Marsh and Kevin Done, 'Hungary: Tough reforms bring rewards', *Financial Times,* December 16, 1996, Survey, p. I.
37. Tony Judt, *A Grand Illusion? An Essay on Europe.* Cited in Jane Perlez, 'New

Bricks, Same Old Walls For Europe's Poor Nations', *The New York Times*, January 24, 1997, p. A2.

38. Anders Aslund, 'Behind the new iron curtain of Europe', *Financial Times*, January 23, 1996, p. 15.
39. Matthew Brzezinski, 'Here Success is Spelled "Dnepropetrovsk"', *The Wall Street Journal*, January 28, 1997, p. A13; Matthew Kaminski, 'Ukraine held hostage by the class of its clans', *Financial Times*, August 7, 1996, p. 2.
40. Avraham Shama, 'Inside Russia's True Economy', *Foreign Policy*, 103 (Spring 1996), p. 112.
41. See Richard Layard and John Parker, *The Coming Russian Boom*. Cited in David Remnick, 'Can Russia Change?' *Foreign Affairs*, 76:1 (January/February 1997), p. 48.
42. 'Transition Report Update', (London: European Bank for Reconstruction and Development), April 1996.
43. The author includes the EMU in the *acquis communautaire*. Although monetary union is part of the Maastricht Treaty, it does not seem absolutely necessary to the future of the EU and the integration of Europe. Niblett, p. 105.
44. Valéry Giscard d'Estaing, 'Europe: Les Raisons de l'Echec' ('Europe: The Reasons for Failure'), *Le Figaro*, January 10, 1995, p. 6. Cited in Niblett, p. 91.

Chapter 7 Japan Stalled

1. Robert Neff, 'Unlocking Japan – At Last?' *Business Week*, April 14, 1997, p. 57.
2. Milton Ezrati, 'Japan's Aging Economics', *Foreign Affairs*, 76:3 (May/June 1997), pp. 96–104.
3. Charles Horner, 'The Third Side of the Triangle: The China–Japan Dimension', *The National Interest*, 46 (Winter 1996/97), pp. 23–31.
4. Brian Beedham, 'Tomorrow's Japan', *The Economist*, July 13, 1996, Survey, pp. 5–6.
5. Karl Schoenberger, 'Has Japan Changed', *Fortune*, August 19, 1996, p. 74; Neff, p. 56.
6. Neff, p. 56.
7. Schoenberger, p. 74.
8. Neff, p. 57.
9. Lionel Barber, 'A deeper transformation', *Financial Times*, April 10, 1996, p. 14.
10. Shoichiro Toyoda, 'Dismantle Japan Inc.', *The New York Times*, April 17, 1997, p. A21.
11. 'Deregulation in Japan: Unwinding red tape', *The Economist*, April 12, 1997, p. 64.

12. Toyoda, p. A21.
13. Peter Passell, 'Economic Scene', *The New York Times*, July 11, 1996, p. C2. Passell cites Alan Feldman of Salomon Brothers and Gary Saxonhouse of the University of Michigan.
14. Ezrati, pp. 96–104.
15. 'China's new tipple', *The Economist*, April 5, 1997, p. 37.
16. 'Cyber-Mogul', *Business Week*, August 12, 1996, p. 56.
17. Peter Landers, 'Media: Second Thoughts', *Far Eastern Economic Review*, March 13, 1997, p. 56.
18. Norihiko Shirouzu and David P. Hamilton, 'Softbank's Buying Spree May Be Hard Act to Follow', *The Wall Street Journal*, August 19, 1996, p. B4.
19. Andrew Pollack, 'Eyes on Higher Things And on the Bottom Line', *The New York Times*, April 2, 1997, p. C1.
20. Pollack, p. C6.
21. Nicholas D. Kristof, 'The Land of Laureates: Japan's Passion is Poetry', *The New York Times*, January 20, 1996, p. 4.
22. Emily Thornton, 'Cars: Seeking Immortality', *Far Eastern Economic Review*, January 30, 1997, p. 50.
23. Chris Gay, 'Policies: To Have and to Hold', *Far Eastern Economic Review*, March 13, 1997, p. 57.
24. Andrew Shipley, 'Japan's Future Jobless', *Far Eastern Economic Review*, April 10, 1997, p. 34.
25. William Dawkins, 'A fresh outlook at the top', *Financial Times*, July 22, 1996, p. 12.
26. Dawkins, p. 12.
27. Dawkins, p. 12.

Chapter 8 China Realism

1. Dominic Ziegler, 'China', *The Economist*, March 8, 1997, Survey, p. 4.
2. 'And never the twain shall meet ...', *The Economist*, March 29, 1997, p. 67.
3. Alastair Iain Johnston, *Cultural Realism: Strategic Culture and Grand Strategy in Chinese History* (Princeton, NJ: Princeton University Press, 1997). See also George Melloan, 'China's Balance of Power Politics in Asia', *The Wall Street Journal*, January 20, 1997, p. A15.
4. Ziegler, p. 4.
5. Ziegler, p. 15.
6. Kathy Chen, 'Chinese President Ratchets Up Reforms', *The Wall Street Journal*, April 7, 1997, p. A11.

7. Joyce Baranthan and Dexter Roberts, 'China: "Grasp the Big Release the Small"', *Business Week*, April 21, 1997, p. 54.
8. Tony Walker, 'Beijing Takes Unsteady Aim at State-owned Industries', *Financial Times*, March 5, 1997, p. 4.
9. Roderick MacFarquhar, 'Demolition Man', *The New York Review of Books*, March 27, 1997, pp. 14, 15 and 17.
10. 'Chinese Firewall: Beijing Seeks to Build Version of the Internet that Can Be Censored', *The Wall Street Journal*, January 31, 1996, p. A1.
11. Nayan Chanda and Karl Huus, 'China: The New Nationalism', *Far Eastern Economic Review*, November 9, 1995, pp. 22–3.
12. John H. Dunning, 'Reappraising the Eclectic Paradigm in an Age of Alliance Capitalism', *Journal of International Business Studies*, 26:3 (Third Quarter 1995), pp. 461–91.
13. Adopted from Dunning, pp. 473–480.
14. Joseph Kahn, 'Chinese Corporations Bulk Up to Take On the World', *The Wall Street Journal*, July 5, 1995, p. A6.
15. Bruce Gilley, 'Great Leap Southward', *Far Eastern Economic Review*, November 23, 1995, pp. 61–2.
16. Seth Faison, 'To Taste Market Woes in China, Try Tsingtao', *The New York Times*, December 25, 1995, pp. C1 and C4.
17. Faison, p. C4.
18. Henry Sender, 'Rat Race: Is it time to get back into Asian equities?', *Far Eastern Economic Review*, January 11, 1996, p. 43.
19. Jeffrey Parker, 'State-sector losses vex Chinese foreign experts', <C-reuters@clarinet.com>, June 11, 1995, (Accessed January 25, 1996).
20. 'Credit in China: Neither Borrower Nor Lender Be', *The Economist*, April 29, 1995, p. 86.
21. Joseph Kahn, 'Private Takeover: Spreading Capitalism, New Entrepreneurs are Remaking China', *The Wall Street Journal*, July 20, 1995, pp. A1 and A10.
22. Joseph Kahn, p. A10.
23. Andrew Tanzer, 'Capitalist pigs', *Forbes*, January 1, 1996, p. 54.
24. 'Investment Opportunities in China', <http://networth.galt.com/guinness/opportunities.htm>, no date, (Accessed January 25, 1996).
25. World Bank, *World Development Report 1994*, (New York: Oxford University Press, 1994). Cited by Yanrui Wu, 'Productivity Growth, Technological Progress, and Technical Efficiency Change in China: A Three-Sector Analysis', *Journal of Comparative Economics*, 21:2 (October 1995), p. 207.
26. Wu, p. 215.
27. Wu, p. 221.
28. Nicholas R. Lardy, 'Economic Engine? Foreign Trade and Investment in China', *The Brookings Review*, 14:1 (Winter 1996), p. 15.

29. Joseph Kahn, 'Surveys Show Profits in China Not So Elusive', *The Wall Street Journal*, September 15, 1995, p. A6.
30. Karl Schoenberger, 'Companies: Motorola Bets Big on China', *Fortune*, May 27, 1996, p. 118.
31. Schoenberger, p 118.
32. Schoenberger, p. 120.
33. Schoenberger, p. 118.
34. 'Technology transfer and China and AT&T', <http://kafka.uvic.ca/tmbclarke/sections/tekchina.html>, no date, (Accessed June 20, 1996).
35. 'AT&T's recent activities in China', <http://kafka.uvic.ca/Ôbclarke/sections/AT&TChina.html>, no date, (Accessed June 20, 1996).
36. 'AT&T, China sign technology transfer and manufacturing deals', <http://www.alt.com/press/0895/950803.cib.html>, no date, (Accessed June 20, 1996).
37. 'China: Microsoft's Long March', *Business Week*, June 24, 1996, p. 52.
38. 'China: Microsoft's ...', p. 53.
39. 'China: Microsoft's ...', p. 54.
40. Tony Walker, 'Chinese lesson', *Financial Times*, February 23, 1995, p. 14.
41. 'And never the twain shall meet ...', *The Economist*, March 29, 1997, p. 67.
42. 'And never ...', p. 67.
43. 'And never ...', p. 67.
44. Tony Walker, 'Living with competition in open-door China', *Financial Times*, April 30, 1996, p. 8.
45. 'Window shopping in Shanghai', *The Economist*, January 25, 1997, p. 62.

Chapter 9 Paradigms for South East Asia

1. Andrew Adonis, 'Determined trend towards Asian values', *Financial Times*, February 24, 1995, p. VIII.
2. 'SE Asia Seeks Flexible Plan on Investment', *Financial Times*, July 10, 1996, p. 4.
3. Peter F. Drucker, 'The Network Society', *The Wall Street Journal*, March 29, 1995, p. A14.
4. Geert Hofstede, *Cultures and organizations: Software of the mind* (London: McGraw-Hill, 1991). Cited by Ryh-song Yeh and John J. Lawrence, 'Individualism and Confucian Dynamism: A Note on Hofstede's Cultural Root to Economic Growth', *Journal of International Business Studies*, 26:3 (Third Quarter 1995), p. 656.
5. Adonis, p. VIII.
6. Grace Goodell, 'Another Way to Skin a Cat', *The National Interest*, 42 (Winter 1995/96), pp. 66–71.

7. 'South-East Asia's octopuses', *The Economist*, July 17, 1993, p. 61.
8. Edward Luce, 'Philippine Factor', *Financial Times*, April 28, 1995, p. 10.
9. 'Asia's New Giants', *Business Week*, November 27, 1995, p. 65.
10. Jung Hoon Derick Sohn, 'Social Knowledge as a Control System: A Proposition and Evidence from the Japanese FDI Behavior', *Journal of International Business Studies*, 25:2 (Second Quarter 1994), pp. 295–324.
11. See John Naisbitt, *Megatrends Asia* (London: Nicholas Brealy Publishing, 1995), and Jim Rohwer, *Asia Rising* (Singapore: Butterworth-Heinemann, 1995). Both were reviewed in the *Far Eastern Economic Review* on November 16, 1995.
12. 'From Strength to Strength – Asean Online', <http://www.indobiz.com/asean/index.html> (Accessed January 10, 1996).
13. Murray Hiebert, 'Still Going Strong', *Far Eastern Economic Review*, December 14, 1995, p. 69.
14. Peter Montagnon, 'Malaysian PM calls for export drive to overcome trade', *Financial Times*, <http://www.usa.ft.com/news/asia-pac/>, January 12, 1996, (Accessed January 13, 1996).
15. K. T. Arasu, 'Indonesia Seen Walking Economic Tightrope', <Rindonesia-economyURGwe_6J5@clari.net> January 5, 1995 (Accessed January 11, 1996).
16. Ted Bardacke, 'Hot Debates on Overheating', *Financial Times*, December 14, 1995, p. 11.
17. Paul Sherer, 'Battered in Bangkok: Thais Face a Reckoning As Their Boom Goes Bust', *The Wall Street Journal*, April 10, 1997, p. A10.
18. Philip Shenon, 'Once an Economic Laggard, Philippines Signals New Life', *The New York Times*, May 7, 1995, p. 4.
19. 'Banking: For Asia's Bad Banks, It's Pay-Up Time', *Business Week*, February 24, 1997, pp. 111–14.
20. David Wessel, 'Capital Flow to Developing Nations Surges 20%', *The Wall Street Journal*, March 24, 1997, p. B9.
21. 'The Asian Miracle: Is it over?' *The Economist*, March 1, 1997, p. 25.
22. Charles Wolf, 'Asia in 2015', *The Wall Street Journal*, March 20, 1997, p. A16; Walter Russell Mead, 'East Asia Needs a New Growth Strategy', *The Wall Street Journal*, April 17, 1997, p. A22.
23. Kieran Cooks and Peter Montagnon, 'A Liberal Pinch of Eastern Spice', *Financial Times*, October 18, 1995, p, 12,
24. Bardacke, p. 11.
25. 'Big Emerging Markets', US Department of Commerce, <http://www.stat-usa.gov/bems/bemsase/bemsase.html> (Accessed January 10, 1996).
26. Thomas J. Duesterberg, 'Trade, Investment, and Engagement in the U.S.-East Asian Relationship', *The Washington Quarterly*, 17:1 (1993), p. 87.
27. Nigel Holloway, 'Now's Your Chance', *Far Eastern Economic Review*, January

30, 1997, pp. 48–9.

28. Edward Luce, 'SE Asia: singularly different', *Financial Times*, December 4, 1995, p. 10.
29. Luce, p. 10.
30. Roy C. Smith and Ingo Walter, 'Rethinking Emerging Markets', *The Washington Quarterly*, 19:1 (Winter 1996), p. 47. The term 'Washington Consensus' was coined by John Williamson of the Institute of International Economics, a think-tank in Washington, DC.
31. Smith and Walter, p. 61.
32. Gerald Segal, '"Asianism" and Asian Security', *The National Interest*, 42 (Winter 1995/96), p. 59.
33. 'Japan's Role in Asia–Pacific Regional Cooperation', <http://apec.tokio.co.jp/apec/../role/index.html>, July 28, 1995 (Accessed January 10, 1996).
34. 'SE Asia seeks flexible plan on investment', p. 4.
35. 'Asian bond markets: First rate', *The Economist*, December 24, 1994–January 6, 1995, p. 95.
36. Conner Middelmann, 'Step closer to becoming a well-rounded market', *Financial Times*, November 16, 1995, p. VI.
37. Asian Fund Management: The Rivals', *The Economist*, November 2, 1995, p. 83.
38. Bruce Gilley and Murray Hiebert, 'The Lions' Share', *Far Eastern Economic Review*, November 9, 1995, p. 76.
39. 'The world's emerging markets all at sea', *The Economist*, January 28, 1995, p. 68.
40. 'The world's emerging markets all at sea', p. 68.
41. 'A question of balance', *The Economist*, April 22, 1995, p. 21.
42. Michiyo Nakamoto, 'Signs of ebbing strength at home', *Financial Times*, March 1, 1995, p. 13.
43. 'JETRO Releases 8th Survey of Japanese-affiliated Makers in Asean', <http:www.jetro.jp/JETROINFO/Press/3-23.html>, March 23, 1995 (Accessed January 10, 1996).
44. 'Japan's New Identity', *Business Week*, April 10, 1995, pp. 108–19.
45. 'JETRO Releases 8th Survey ...'
46. Nakamoto, p. 13.
47. Dan Biers, 'Japan's Asian Neighbors, in a Shift, Are Tuning In to Its Cultural Exports', *The Wall Street Journal*, October 25, 1994, p. A12.
48. H.H. Aswicahyono and Hal Hill, 'Determinants of Foreign Ownership in LDC Manufacturing: An Indonesian Case Study', *Journal of International Business Studies*, 26:1 (First Quarter 1995), p. 140.
49. Aswicahyono and Hill, p. 141.
50. Nigel Holloway, 'Level Pegging', *Far Eastern Economic Review*, September 28, 1995, pp. 100 and 102.

51. Aswicahyono and Hill, pp. 142–5 and 148–50.

Conclusions

1. Dani Rodrik, *Has Globalization Gone Too Far?* (Washington, DC: Institute for International Economics, 1997), p. 83.
2. Robert Kuttner, 'Owning Up to the Costs of Free Trade', *Business Week*, April 28, 1997, p. 22.
3. Douglas Lamont, 'A Winning Strategy', *Asia Wall Street Journal*, April 11, 1997, p. 7.

Index